Clinical Aspects of Endotoxin Shock

Handbook of Endotoxin

Series Editor: R.A. PROCTOR

VOLUME 4

Clinical Aspects of Endotoxin Shock

Editor:

R.A. PROCTOR

University of Wisconsin Medical School,
Madison, Wisconsin, U.S.A.

1986

ELSEVIER

Amsterdam – New York – Oxford

ISBN 0 444 90390 9

Library of Congress Cataloging-in-Publication Data
Main entry under title:

Clinical aspects of endotoxin shock.
   (Handboek of endotoxin ; v. 4)
   Includes bibliographies and index.
   1. Septic shock. 2. Endotoxins--Physiological
effect. I. Proctor, R. A. (Richard Allan), 1945-
II. Series. [DNLM: 1. Endotoxins. 2. Shock, Septic.
QA 140 C641]
RC182.S4C57 1986   615.9'5299   85-29273
ISBN 0-444-90390-9 (U.S.)

Published by:
Elsevier Science Publishers B.V.
P.O. Box 1126
1000 BC Amsterdam

Sole distributors for the USA and Canada:
Elsevier Science Publishing Co. Inc.
52 Vanderbilt Avenue
New York, NY 10017

Typeset in The Netherlands by Pecasse Intercontinental B.V., Maastricht
Printed in The Netherlands by Casparie, Amsterdam

Contributors

DR. L.T. ARCHER
Veterans Administration Medical
  Center
Research 151-C
921 N.E. 13th Street
Oklahoma City, OK 73104
U.S.A.

DR. W. CRAIG
Department of Medicine
University of Wisconsin Medical
  School
600 Highland Avenue
Madison, WI 53706
U.S.A.

DR. R.H. DEMLING
Longwood Area Trauma Center
75 Francis Street
Boston, MA 02115
U.S.A.

DR. S. GUDMUNDSSON
Hvaffaleiti 113
108 Reykjavik
Iceland

DR. R.M. GUNNAR
Department of Medicine
Loyola University Stritch
  School of Medicine
2160 South First Avenue
Maywood, IL 60153
U.S.A.

DR. D. HALE
Department of Medicine
Loyola University Stritch
  School of Medicine
2160 South First Avenue
Maywood, IL 60153
U.S.A.

DR. R.J. HAMILL
Department of Medicine
Veterans Administration
  Medical Center
2002 Holcumbe Boulevard
Houston, TX 77211
U.S.A.

DR. J.H. JORGENSEN
Department of Pathology
University of Texas
  Health Sciences Center
7703 Floyd Curl Drive
San Antonio, TX 78284
U.S.A.

DR. H.S. LOEB
Department of Medicine
Loyola University Stritch
  School of Medicine
2160 South First Avenue
Maywood, IL 60153
U.S.A.

DR. D.G. MAKI
Department of Medicine
University of Wisconsin
  Hospital and Clinics
600 Highland Avenue
Madison, WI 53792
U.S.A.

DR. R.A. PROCTOR
Department of Medical
  Microbiology
University of Wisconsin
  Medical School
1300 University Avenue
Madison, WI 53706
U.S.A.

DR. J.A. ROBINSON
Section of Clinical Immunology
  and Rheumatology
Loyola University Stritch
  School of Medicine
2160 South First Avenue
Maywood, IL 60153
U.S.A.

DR. L.S. YOUNG
Kuzell Institute for Arthritis
  and Infectious Diseases
Division of Infectious Diseases
Pacific Presbyterian Medical
  Center
2200 Webster Street
San Francisco, CA 94115
U.S.A.

Contents

Chapter 3
Endotoxin shock in man caused by gram-negative bacilli – etiology, clinical features, diagnosis, natural history and prevention
R.J. Hamill and D.G. Maki

Chapter 4

Clinical applications of the Limulus amebocyte lysate test

J.H. Jorgensen

Contents

Chapter 5
Role of antibody in the prevention and pathogenesis of endotoxin and gram-negative septic shock
R.A. Proctor

Chapter 6
Cardiopulmonary dysfunction from sepsis: diagnosis and treatment
R.H. Demling

Chapter 7
Role of antibiotics in endotoxin shock
S. Gudmundsson and W.A. Craig

Chapter 8
Role of glucocorticoids in the treatment of endotoxin shock
L.S. Young

Introduction

The role of endotoxin in the pathogenesis of gram-negative septic shock and its treatment

The role of endotoxin as a cause for shock during gram-negative sepsis has been a major area of controversy. The striking similarity between the acute pathophysiological manifestations following endotoxin challenge and those seen during gram-negative sepsis led to the assumption that the two processes were identical. The release of endotoxin from gram-negative organisms provided a ready explanation for the connection between endotoxin and gram-negative septic shock. Indeed, an argument can be made that endotoxin on the surface of the bacteria might activate many mediators even if it is not released. The finding that antibody reacts with endotoxin in situ (reviewed in Chapter 5) provides evidence for the availability of endotoxin sites to immunoglobulin. The concept that endotoxin released from bacteria plays a role in the pathogenesis of human gram-negative sepsis was given clinical support by the studies of tetracycline treatment of *Brucella melitensis* infections (Abernathy and Spink 1958; Spink et al 1948). Although the blood was cleared of organisms, several patients treated with chlortetracycline developed acute pyrexia and shock. Recently, Shenep and Morgan (1985) reported that antibiotic therapy of patients with gram-negative sepsis resulted in the release of endotoxin, as measured by the Limulus amebocyte lysate assay. Also, gram-negative organisms unable to survive in the host, for example cryophils which grow only at temperatures well below body temperature or highly serum-sensitive *Escherichia coli*, still caused fever and shock when infused into the host (Borden and Hale 1951; Braude 1980). Finally, the incidence of shock is much higher in gram-negative rod sepsis as compared to gram-positive cocci sepsis (Braude 1980). Although all these findings make a strong case for a role of endotoxin in gram-negative septic shock, questions can be raised (Rietschel et al 1982).

Endotoxin cannot be demonstrated in the blood of all patients with gram-negative sepsis. Jorgensen reviews the relevant data extensively in Chapter 4. Clearly, not all cases of gram-negative sepsis result in endotoxin release. This does not rule out endotoxin as a major factor in many cases of gram-negative sepsis, but suggests that many other toxins may be produced which will be important to the final outcome in the host. Studies using newer Limulus amebocyte lysate methodologies and testing for multiple toxins may help us to better understand the cases in which endotoxin is playing a determinative role.

Several animal models have raised questions about the importance of endotoxin. Quarles et al (1974) placed a semipermeable membrane in the vascular tree of goats. As the number of *Serratia marcescens* increased within the chamber, the goats became toxic. Because the membrane would only permit passage of molecules smaller than 15000 daltons and because endotoxin molecules were felt to be much larger, the conclusion was that endotoxin could not be causing the reaction. However, recent studies have shown that biologically active subunits of endotoxin could readily traverse the membrane (Chapter 7, by C.R.H. Raetz in Volume 1), thus endotoxin may well have been contributing to the response.

Another animal model presented the following paradox: C3H/HeJ mice which were highly resistant to endotoxin challenge were very susceptible to *Salmonella typhimurium* (O'Brien et al 1980). This would seem to imply that the toxic actions of endotoxin could be differentiated from the lethality of *S. typhimurium*. However, the chemotactic and mitogenic effects of endotoxin are greatly decreased in the C3H/HeJ mice (reviewed by D.L. Rosenstreich in Chapter 4, Volume 3). Moreover, at least 3 genes control mouse susceptibility to *S. typhimurium* (O'Brien et al 1980). Thus, the interactions between C3H/HeJ mice and *S. typhimurium* are complex and it would be an oversimplified view to rule out an important role for endotoxin in mouse typhoid.

Similarly, studies performed in mice rendered highly sensitive to endotoxin by pretreatment with *Bacillus Calmette-Guérin* (BCG) were found to die with viable *S. typhimurium* counts identical to those found in untreated mice (Senterfitt and Shands 1968). Because BCG also activates macrophages, the interactions between mice, endotoxin, and *S. typhimurium* are also complex. Thus, one might find both protection and sensitization, with a resultant balancing of BCG effects on the ultimate outcome.

Human volunteers and rabbits made tolerant to the pyrogenic effects of endotoxin still responded with fever to *S. typhimurium*, *Francisella tularensis*, and *E. coli* infections (Rietschel et al 1982). This suggests that the sustained fever seen in these infections does not come from endotoxin alone, but it does not rule out a role for endotoxin in the pathogenesis of these infections.

A multiplicity of products released during a bacterial infection acts on macrophages to produce endogenous pyrogen (Dinarello and Wolff 1978) and accounts for the fevers seen in endotoxin-tolerant hosts. Therefore, although the animal models raise some questions, the data presented in these four volumes strongly favor a role for endotoxin in the pathogenesis of gram-negative infections. The human pathophysiologic responses to gram-negative sepsis are reviewed in Chapter 1. In Chapter 2, Dr Archer reviews the pathologic manifestations of gram-negative sepsis and endotoxin. Surprisingly, reviews of the histologic manifestations of endotoxin are rarely found. The clinical setting for sepsis and endotoxemia are reviewed in Chapter 3 of this volume by Drs. Richard Hamill and Dennis Maki. The role of antibiotics, pressors, colloids, anti-endotoxin antibody, and glucocorticoids in the treatment of gram-negative sepsis and shock are critically reviewed in Chapters 4–8. In each of these chapters, the benefits of each of these therapeutic modalities are reviewed, controversies presented, and future directions indicated. Clearly, antibiotics, colloids and pressors are well established as beneficial. The studies to date suggest that anti-endotoxin antibodies will be very important in the prevention and treatment of endotoxin shock. Finally, the review of glucocorticoid therapy for endotoxin shock points out the need for well-controlled, prospective studies.

I would like to thank all the contributors to Volumes 1-4. The excellence of this series reflects their efforts and the efforts of the editors of Volumes 1-3, Drs Rietschel, Hinshaw and Berry. I would also like to thank three of my previous professors: Dr Robert Fekety for stimulating my interest in infectious diseases and Drs Calvin Kunin and Joe Wilson for their valuable criticism and suggestions for my research work on endotoxin.

RICHARD A. PROCTOR

REFERENCES

Abernathy RS, Spink WW (1958) Studies with brucella endotoxin in humans. The significance of susceptibility to endotoxin in the pathogenesis of brucellosis. *J. Clin. Invest. 37*, 219–231.

Borden C, Hale W (1951) Fatal transfusion reactions from massive bacterial contamination of blood. *N. Engl. J. Med. 245*, 760–763.

Braude AI (1980) Endotoxic immunity. *Adv. Intern. Med. 26*, 427–445.

Dinarello CA, Wolff SM (1978) Pathogenesis of fever in man. *N. Engl. J. Med. 298*, 607–612.

O'Brien AD, Rosenstreich DL, Scher I, Campbell GH, MacDermott RP, Formal SB (1980) Genetic control of susceptibility to *Salmonella typhimurium* in mice: role of the Lps gene. *J. Immunol. 124*, 20–24.

Quarles JM, Belding RC, Beaman TC, Gerhardt P (1974) Hemodialysis cultures of *Serratia marcenscens* in a goat-artificial kidney-fermentor system. *Infect. Immun. 9*, 550–558.

Rietschel ET, Schade U, Jensen M, Wollenweber HW, Lüderitz O, Greisman SG (1982) Bacterial endotoxins: chemical structure, biological activity and role in septicaemia. *Scand. J. Inf. Dis., Suppl. 31*, 8–21.

Senterfitt VC, Shands JW Jr (1968) Salmonellosis in mice infected with *Mycobacterium tuberculosis* BCG I. Role of endotoxin in infection. *J. Bacteriol. 96*, 287–292.

Shenep JL, Barton RP, Morgan KA (1985) Role of antibiotic class in the rate of liberation of endotoxin during antibiotic therapy for experimental gram-negative sepsis. *J. Infect. Dis. 151*, 1012–1018.

Spink W, Braude A, Castaneda MR, Goytia R (1948) Aureomycin therapy in human Brucellosis due to *Brucella melitensis. J. Am. Med. Assoc. 138*, 1145–1152.

Handbook of Endotoxin, Vol. 4: Clinical Aspects of Endotoxin Shock
R.A. Proctor, editor
© Elsevier Science Publishers B.V., 1986

CHAPTER 1

Pathophysiology of endotoxin shock in man

DAVID J. HALE, JOHN A. ROBINSON, HENRY S. LOEB AND ROLF M. GUNNAR

1. INTRODUCTION

The shock syndrome in man may be defined as a profound alteration in physiologic regulation of the circulation resulting in inadequate tissue oxygenation and anaerobic metabolism, frequently terminating in death. When the syndrome is associated with infection, the term septic shock is used.

Septic shock in man is often inaccurately equated with gram-negative bacteremic shock (GNS), but septic shock may be associated with infection by all classes of microorganisms including bacteria, viruses, fungi, protozoa, spirochetes, and rickettsia. Toxic shock syndrome, a syndrome associated with gram-positive infection, should serve as a current reminder of the clinical importance of septic shock unrelated to gram-negative infection. However, gram-negative bacilli are the most common cause of nosocomial bacteremia in the United States and the most common cause of septic shock (McCabe 1974). Furthermore, shock induced by the injection of endotoxin in laboratory animals produces pathophysiologic alterations similar to GNS in humans. While differences between experimental bacteremic shock and experimental endotoxin shock are discussed in Chapter 2, recent evidence suggests that the shock syndrome associated with gram-negative bacteremia in man is most likely caused by endotoxin (Ziegler et al 1982).

This chapter will review the pathophysiologic mechanisms responsible for the profound alterations in systemic vascular resistance, myocardial contractility, pulmonary function, coagulation, and cellular oxygen extraction that are associated with GNS in adults. We will also discuss, especially in terms of the pathophysiologic insights they provide, two potentially useful modes of therapy: opiate antagonists and endotoxin antibodies.

2. CLINICAL FEATURES

2.1. Epidemiology

GNS is apparently a relatively modern syndrome. Clinical descriptions of it occurred sporadically in the literature before 1951 (Waisbren 1951). Since then, many clinical centers, including our own (Loeb et al 1967; Winslow et al 1973), have reported their experience with GNS. Mortality data from some of these reports are summarized in Table 1 (Dupont and Spink 1969; Loeb et al 1967; Nishijima et al 1973; Waisbren 1951; Winslow et al 1973; Ziegler et al 1982). All series have several features in common. The patients affected have compromised host defenses resulting from their underlying disease process, medical therapy, indwelling catheters, protein-calorie malnutrition or advanced age. Furthermore, the risk of death from GNS is clearly related to the severity of the underlying disease process. McCabe and Jackson (1962) found that those patients classified as having a rapidly fatal underlying disease experienced a 91% mortality rate; patients with nonfatal underlying disease had an 11% mortality rate. During the ensuing 20 years, other investigators also found that the mortality of GNS increases as the severity of the underlying disease status of the host increases (Dupont and

TABLE 1 *Mortality of gram-negative bacteremia (GNB) and gram-negative bacteremic shock (GNS)**

Clinical series		GNB		GNS	
		No. of patients	Mortality rate (%)	No. of patients	Mortality rate (%)
Waisbren	1951	31	16	15	27
Loeb et al	1967	–	–	7	57
DuPont and Spink	1969	655	50	265	82
Winslow et al	1973	–	–	31	65
Nishijima et al	1973	–	–	32	63
Ziegler et al	1982	212	31	71	61
Total		898	32	421	59

*Pediatric population excluded

Spink 1969; Kreger et al 1980; Weisel et al 1977; Winslow et al 1973; Ziegler et al 1982).

In spite of our awareness of the syndrome of GNS and the availability of effective antibiotics, the incidence of GNS is progressively increasing, and mortality is not decreasing. The increase is attributable to changes in medical care modalities including chemotherapy, radiation therapy, invasive instrumentation, and an increasingly elderly hospital population. Fundamentally, GNS is a nosocomial, iatrogenic problem.

The dividing line between gram-negative bacteremia (GNB) and gram-negative bacteremic shock (GNS) is somewhat arbitrary, with GNB being somewhat less lethal than GNS (see Table 1). It is difficult to compare the experience of one clinical center with another because of differences in the definition of GNS, patient populations, and decade of study. When the data from the selected series in Table 1 are combined and weighted equally, the overall mortality of GNB is 37% while the mortality of those developing GNS is 59%. It is apparent that the mortality of established GNS has not decreased appreciably since the original description (Waisbren 1951).

McCabe has estimated the incidence of GNB to have increased 20 times in a 20-year period as a result of modern medical practices introduced during that period (McCabe 1974). He estimates the annual incidence of GNB in the United States to be 330 000 cases, or 1% of hospitalized patients, with an estimated annual mortality of 132 000. An estimate from the Center of Disease Control is somewhat more conservative with an annual incidence of 70 000 cases of GNB per year with an estimated mortality of 25% (Wolff and Bennett 1974).

2.2. Signs and symptoms of gram-negative shock

The syndrome of GNS may develop insidiously over several hours or rapidly in minutes. The typical patient is gravely ill from a noninfectious disease. Host defense may be further compromised with clinical instrumentation such as a Foley catheter, or a recent surgical procedure such as cystoscopy or colonic surgery. Often the first change noted is unexplained tachypnea followed by fever, chills, and hypotension, not necessarily in that order. The patient may be warm peripherally in spite of hypotension during the early phase of the syndrome.

Altered mental status and reduced urine output ensue. Leukopenia may be detected initially followed by leukocytosis. Respiratory alkalosis will be detected initially followed later by hypoxia and metabolic acidosis. The chest X-ray may show alveolar infiltrates. Blood cultures will grow gram-negative bacilli corresponding to the flora commonly found in the organ system that is the fount of sepsis.

3. HEMODYNAMICS OF GRAM-NEGATIVE SHOCK

3.1. Systemic vascular resistance

Waisbren's report of GNS was purely descriptive. The only quantitative hemodynamic measurements available to him were systemic blood pressure measurements by sphygmomanometry and heart rate. A need for more precise means of accurately categorizing degrees of GNS and response to therapy was a strong evolutionary force for the implementation of invasive hemodynamic instrumentation in clinical shock research units. The measurements of arterial pressure, central venous pressure, cardiac output, pulmonary arterial pressure, and left ventricular pressure became commonplace in clinical centers investigating shock (Table 1). An important byproduct of that research effort has been the development of hemodynamic monitoring techniques now commonly employed in practically every intensive care unit nationwide.

During the past 20 years, we have learned that the hemodynamics of the early phases of the GNS syndrome are characterized by decreased systemic vascular resistance (SVR) with a compensatory increase in cardiac output

TABLE 2 *Hemodynamics of gram-negative bacteremia and shock*

Clinical series		Systemic vascular resistance	Myocardial function
Udhoji and Weil	1965	↓ and ↑	↓
Loeb et al	1967	↑	↓
MacLean et al	1967	↓ and ↑	↓
Siegel et al	1967	↓	↓
Kwaan and Weil	1969	↑	↓
Winslow et al	1973	↓	↓
Nishijima et al	1973	↓ and ↑	↓
Weisel et al	1977	–	↓

(CO). As the time course of the GNS syndrome progresses, CO decreases and a compensatory increase in SVR is observed (Gunnar et al 1973). However, the hemodynamics of GNS could not be concisely characterized until more than 20 years of clinical observation had been recorded.

Table 2 shows that many early clinical investigators found both increased and decreased SVR associated with GNS. Siegel (Siegel et al 1967) first noted that all patients had a relative decrease in what he termed 'net vascular tone'. We suspected that the disparity in observations of SVR could be explained by examining the time course of hemodynamic measurements of each of the series listed in Table 2: decreased SVR is an early manifestation; increased SVR occurs at the end stages.

In order to test the hypothesis that GNB is associated with decreased SVR, we performed a prospective study of the hemodynamics of GNB (Blain et al 1970). To place our model in context, it should be noted that Scott (1929) first emphasized the frequent association of bacteremia with urologic surgical procedures, and this experience has been repeated in a more recent series (Sullivan et al 1973). In fact, approximately one-third of patients undergoing elective urologic surgical procedures develop bacteremia, and this predictable occurrence of sepsis provided an opportunity to test our hypotheses prospectively. We measured hemodynamic parameters including SVR in patients before and after urologic procedures and found that SVR decreased in the bacteremic group (Blain et al 1970).

Experimental work and hypotheses followed as investigations attempted to unravel the mechanism(s) of the paradoxical decrease in SVR observed in the presence of hypotension during the early phase of GNB. Hinshaw found that endotoxin is capable of activating an endogenous substance in the blood which alters pulmonary vascular resistance in an isolated canine lung preparation (Hinshaw et al 1957). Circulating vasoactive substances such as histamine, serotonin, prostaglandins, bradykinin, and complement are all capable of inducing the changes in SVR associated with GNS syndrome. Nies et al (1968) showed that bradykinin, a potent vasoactive peptide, appears in the circulation of monkeys given endotoxin, and the appearance of bradykinin is associated with hypotension. Mason and Melmon demonstrated a decreased peripheral resistance in an occluded forearm preparation in man during kinin infusion (Mason and Melmon 1965).

It is tempting to attribute the changes in SVR observed during GNS to activation of the kinin system by circulating endotoxin. Endotoxin activates Hageman factor in the coagulation cascade, and activated Hageman factor in turn is capable of liberating kinins by activating prekallikrein. However, kinins in man are rapidly neutralized by kininases and are unlikely to fully explain the profound, prolonged decrease in SVR associated with GNS.

Our attention turned to the complement system (Robinson et al 1975). Bokisch et al (1973) reported that complement was activated in dengue hemorrhagic shock. Robinson et al (1975), using the prospective urologic GNB model mentioned earlier (Blain et al 1970), found that patients developing endotoxemia subsequent to genitourinary surgery activated the complement system and converted prekallikrein to kinin while experiencing a transient decrease in SVR.

3.1.1. *Activation of complement cascade during gram-negative shock*

The complement system is an extremely complex mosaic of more than 18 proteins of disparate biologic activities. Activation of the system can be conceptually viewed as a step-wise activation of proteolytic precursors that cleave off biologically active fragments at multiple steps in the cascade sequence. These products have diverse biologic effects on the microvasculature, inflammatory cells, immune response cells and endothelium, with a common goal of providing an optimal inflammatory response. Some of the biologic functions of complement that may be directly relevant to the sepsis syndrome include production of anaphylatoxins with microvascular and permeability effects, histamine release, hypercoagulant states, platelet aggregation, exudation of proteins, microvascular fluid shifts and lipopolysaccharide detoxification.

There are two entry pathways into the terminal sequence or biologically active portion of the complement cascade system. The classical pathway is denoted as such because of the early demonstration that an immune complex was required to initiate C1 esterase formation and subsequent assembly of $C\overline{42}$ to a C3 convertase. Regulation of this portion of the pathway is mediated by an inhibitor to C1, a C4 binding protein and C3b inactivator (C3bINA). The other entry route is more primitive than the classical pathway and is designated the alternative pathway. This pathway serves a primary recognition defense role since it can be nonimmunologically activated by polysaccharides, lipid A, other cell wall substances and aggregated antibodies bypassing the mandate for an antecedent specific immune response. Activation of the alternative pathway is maximally effective when driven by a positive feedback or amplification loop that requires the presence of the C3 cleavage fragment C3b. Alternative pathway factors B, D and properdin have properties quite similar to C1, 4 and 2 in that they assemble and stabilize a convertase for C3 and C5. Regulation of the alternative pathway amplification loop is provided by at least two proteins, C3bINA and β-1H-globulin. Both dampen C3b-enhancing activities in the amplification loop and act as regulatory or check-point mechanisms that prevent uncontrollable cascade type proteolysis and its potentially severe biologic consequences.

A large body of experimental evidence garnered both in vitro and in higher primates, including man, provides an inescapable link between GNB, endotoxemia, and shock. Recent awareness that there are multiple immunologic and nonimmunologic entry points into the biologically active portion of the complement pathway provide an ample theoretical basis for many of the clinically observed hemodynamic and coagulation phenomena consistently observed in GNS.

As with all simplistic biologic hypotheses, however, as additional data became available, the proposed model became quite a bit more complex. Early observations, based on crude but the only extant measurements available, clearly showed that GNB or endotoxemia leads to predictable decreases in SVR (Winslow et al 1973) and that this hemodynamic event is associated with the transient emergence of plasma bradykinin that had been generated by the activation of the kallikrein system (Nies et al 1968). The in vitro observations that LPS induced C3 cleavage buttressed the earlier observations of reduced serum complement during endotoxemia (Gilbert 1962) and provided the rationale for blunting the hemodynamic effects of intravenous endotoxin by prior in vivo decomplementation (Fong 1970).

Initially, most evidence suggested that the polysaccharides and lipid A moieties of LPS were capable of activating the alternative pathway, and that this was the primary route that subsequently led to C3 and C5 cleavage and generation of the anaphylatoxins C3a and C5a that are potent mediators of increased vascular permeability, fluid shifts, and decreased SVR that lead to the sepsis syndrome. Robinson et al correlated the emergence of Limulus positivity (a biologic assay for endotoxin) in the plasma of patients undergoing transurethral prostatectomy or cystoscopy with decreases in both SVR and an alternative pathway protein (Robinson et al 1975). Fearon detected significant decreases in properdin, factor B, C3 and late complement components in patients with bacteremia and shock (Fearon 1975). Since the development of new and more sensitive assays for the functional activity of this pathway, striking confirmatory evidence of its major involvement has become available. Activation of the alternative pathway produces weak hemolytic activity, and therefore this classic measurement of complement activation is not often helpful. Moreover, individual component assays of factors B, D, and properdin did not consistently predict alternative pathway functional activity. The development of a new functional assay of the alternative pathway confirms that its activation may be markedly depressed but not recognized if only hemolytic assays or specific assays for C3, factor B and D are assessed (Joiner 1983).

In spite of the strong evidence for predominant alternative pathway involvement during endotoxemia, studies in C4-deficient guinea pigs suggested that at least the thrombocytopenia occurring during gram-negative sepsis was being mediated via the classical pathway (Kane 1973). The increased procoagulant activity also occurring during GNS may be due to direct endotoxin stimulation of monocyte procoagulant production (Osterud 1982). It now appears that lipopolysaccharide may directly activate the complement system by inducing C1 assembly in addition to its previously recognized alternative pathway entry point. Additional mechanisms whereby the complement pathway can be activated during sepsis syndromes include the recent recognition that kallikrein can be an effective factor D imitator and cleave factor B to Bb and assemble the C3 convertase C3bBb (DiScipio 1982). This mechanism may act as an important supplemental factor for complement activation during endotoxemia. Although several plasma protease inhibitors regulate kallikrein induction of the C3 convertase, several investigators have shown that prekallikrein is depleted during endotoxemia, presumably secondary to Hageman factor cleavage by lipopolysaccharide (Nies et al 1969; Robinson et al 1975). The subsequent activation of the kallikrein cascade leads to generation of bradykinin, a very transient, rapidly depleted, but potent hypotensive peptide (Harpel 1975). Lastly, plasmin and trypsin may directly activate C1 and C3, but their roles in bacteremia-associated hypotension is unclear at present.

Regardless of the entry point or pathway traveled, once C3 is cleaved, certain inexorable and potentially fatal vasoactive phenomena characteristic of the anaphylatoxins C3a and C5a result, assuring anaphylatoxins certainly a dominant role in GNS. C3a and C5a induce histamine release from mast cells and basophils, platelet release of serotonin and smooth muscle contraction. C5a is a potent chemotactic factor and granulocyte aggregator. The regulation of anaphylatoxic activities is dependent upon the presence of a serum enzyme, carboxypeptidase B, which cleaves arginine from the vasoactive peptides and renders them inactive.

Now that many of the early hemodynamic events of endotoxemia are inextricably connected to the generation of vasoactive phenomena arising from activation of the complement pathway, emphasis may shift to further understanding the role of these relatively newly recognized groups of regulator or control proteins (C3bINA, B1H, etc.) and their ability to modulate the sepsis syndrome. Perhaps GNS only arises in individuals with relative preexisting inhibitor or control protein deficiencies, or is perpetuated by suppression of regulator protein synthesis or action by one of the moieties of endotoxin. Further understanding in this area could lead to important new therapeutic modalities.

In summary, our current, central hypothesis is that endotoxin is capable of activating the complement system in man, and activated complement is responsible for the profound alteration in integrity of the circulatory system resulting in the GNS syndrome.

3.2. Myocardial dysfunction

The profound hemodynamic changes associated with the GNS syndrome have only been quantitatively described during the past 20 years in which methods of invasive, bedside, hemodynamic monitoring have been available. We now suspect that patients described by Waisbren as cool, clammy, hypotensive and lethargic probably had decreased cardiac output and increased SVR. As more experience with accurate hemodynamic measurements was published, two clear conclusions emerged: patients with GNS syndrome develop progressive myocardial dysfunction as the duration of the shock syndrome increases, and the mortality of GNS increases as myocardial function declines.

Each of the clinical series listed in Table 2 concluded that myocardial dysfunction occurs in GNS. Siegel et al (1967) stated that all patients with septic shock have a degree of left ventricular dysfunction. Udhoji and Weil found a decrease in cardiac index to be associated with increased mortality (Udhoji and Weil 1965). MacLean et al (1967) noted that all patients with a decreased cardiac index and acidosis died. Nishijima emphasized the significance of myocardial dysfunction induced by GNS in his review. He concluded that mortality increases as the cardiac index decreases and that the relationship is linear (Nishijima et al 1973).

We studied changes in left ventricular function in response to plasma volume expansion in patients with GNS using each patient as his own control (Winslow et al 1973). Despite marked increases in left ventricular filling pressure in response to plasma volume expansion, left ventricular stroke work and cardiac index failed to rise significantly. From these data, we concluded that myocardial function is impaired in GNS.

More recently, Weisel et al developed serial left ventricular performance curves by plotting left ventricular stroke work index in response to plasma volume expansion in patients with bacteremia in a trauma unit. All patients with bacteremia were found to have myocardial dysfunction. More importantly, myocardial dysfunction was found to be reversible in those patients surviving their bacteremic episode (Weisel et al 1977).

The reversible nature of myocardial dysfunction associated with endotoxin shock is well supported in the experimental literature. Kondo produced myocardial dysfunctions in an experimental endotoxin shock model in dogs, but found that myocardial function normalized when the dysfunctioning hearts were transplanted to dogs not in endotoxin shock (Kondo et al 1975).

Other recently published clinical studies using serial gated blood pool imaging techniques along with hemodynamic measurements have shown that myocardial dysfunction associated with GNS (MD-GNS) may be localized and not global. Furthermore, the occurrence of segmental myocardial dysfunction cannot be attributed to underlying coronary artery disease (Ellrodt et al 1983; Kimchi et al 1983). It is also apparent that MD-GNS occurs without apparent myocardial necrosis. The clinical literature does not contain frequent references to significant electrocardiographic changes or elevation of cardiac enzymes resulting from MD-GNS. It is doubtful that the regular occurrence of myocardial necrosis would have been overlooked in the extensive clinical literature pertaining to GNS published during the past 20 years.

Our experience and interpretation of the extant clinical literature lead us to conclude that myocardial dysfunction is associated with GNS in man, and that myocardial dysfunction is reversible. Furthermore, the functional defect producing myocardial dysfunction may be regional, not global, and these segmental, functional abnormalities occur in the absence of myocardial necrosis.

The pathophysiologic mechanism responsible for the production of MD-GNS is not known. Our current hypothesis is that the activation of vasoactive substances referred to earlier in this chapter (see 3.1.1) produce alterations in effective myocardial blood flow resulting in localized myocardial dysfunction. The reader is referred to the experimental works of Hess for further discussion of potential pathophysiologic mechanisms (Hess and Krause 1981; Hess et al 1981).

4. MISCELLANEOUS PATHOPHYSIOLOGIC ASPECTS OF GRAM-NEGATIVE SHOCK

4.1. Acute respiratory distress

GNS is often heralded by unexplained tachypnea and respiratory alkalosis followed by hypoxia and pulmonary infiltrates. Earlier authors found respiratory insufficiency to be an important cause of death in GNS (Clowes 1970). With modern respiratory support, pulmonary insuffuciency is less likely to be the mode of death in GNS (Vito 1974). However, respiratory distress associated with GNS remains an important clinical problem. In one series, the most common cause of acute respiratory failure on a general surgical service excluding trauma patients was sepsis; conversely, 40% of patients eventually found to be septic presented with respiratory symptoms (Vito et al 1974).

The pathogenesis of the acute respiratory distress syndrome accompanying GNS (ARDS-GNS) remains obscure. The experimental literature relative to this important topic is summarized in Chapters 2 and 6, and in Volume 2 of this series. Autopsy studies of the respiratory distress syndrome in man reveal a predictable pathologic sequence characterized by microemboli, interstitial and alveolar edema initially followed in one or two days by gross alveolar hemorrhage creating the pathologic appearance of liver. If the patient survives 5–7 days, hyaline membranes will appear (Blaisdell and Schlobohm 1973). The problem appears to be initiated by damage to the pulmonary capillaries resulting in alveolar hemorrhage and proteinaceous exudate.

The mechanism producing the loss in pulmonary capillary integrity may be diffuse microembolization (Blaisdell et al 1970). More likely, however, pulmonary capillary permeability is altered by vasoactive substances such as C5a arising secondary to endotoxin activation. The presence of this anaphylatoxin will lead to aggregation of hydrolase-rich neutrophils in the pulmonary vasculature (Jacobs 1981). Hosea et al found that pneumococcal bacteremia was capable of inducing leakage of albumin into the alveoli of guinea pigs, and the pulmonary changes were accompanied by evidence of complement activation. However, complement-depleted guinea pigs did not develop pulmonary lesions (Hosea et al 1980). The authors also concluded that complement activation induced alterations in pulmonary capillary permeability. It is likely that ARDS-GNS may prove to be another manifestation of the central role of complement activation in GNS.

4.2. Disseminated intravascular coagulation

A coagulative disorder associated with infection and characterized by increased fibrin deposition and increased fibrinolysis has been variably termed the defibrination syndrome, consumptive coagulopathy and more commonly, disseminated intravascular coagulation (DIC). The role of endotoxin in the production of DIC has been emphasized since the 'generalized Shwartzman reaction' described fibrin deposition in the glomeruli of rabbits given endotoxin at critically timed intervals. DIC in man associated with GNS may be attributed to activation of Hageman factor by endotoxin. Hageman factor activation results in both fibrin formation and fibrinolysis, both features of DIC.

Whether DIC is a primary event in GNS or results from the increased thrombogenicity accompanying endothelial damage and monocyte procoagulant remains unresolved. In any event, thrombocytopenia is commonly noted and may result from aggregation of platelets resulting from exposure to endotoxin and/or damaged endothelium.

However, DIC rarely produces overt hemorrhage. McCabe estimated the 10% to 20% of patients with GNS will exhibit alterations in hemostasis (McCabe 1974). In our experience, changes in hemostasis resulting from DIC are usually a preterminal event when the GNS syndrome is far advanced. Earlier in the course of GNS, DIC will be detected by in vitro studies demonstrating thrombocytopenia, increased fibrin split products and reduced fibrinogen levels. Clinically significant elevations in prothrombin time and partial thromboplastin time, however, rarely occur, and clinically significant hemorrhage is exceptional.

While these abnormalities detected by in vitro testing offer an opportunity for insight into the pathophysiologic basis of GNS, it is important to note this DIC will usually be reversible if the GNS syndrome responds to therapy.

4.3. Arteriovenous shunting

Siegel noted that arteriovenous oxygen extraction was paradoxically reduced during GNS (Siegel 1967). He termed this problem defective oxygen transport. Initially, the opening of anatomic arteriovenous shunts diverting blood from the capillary bed was proposed as the mechanism responsible for decreased oxygen extraction. However, anatomic shunts have never been convincingly demonstrated.

Functionally, in GNS, blood may pass from the arterial to the venous system without becoming fully desaturated even in the presence of a decreased cardiac output and increased oxygen demand. The cause of the failure to extract oxygen (physiologic shunt) remains obscure but anaerobic metabolism results, producing lactic acidosis and death.

5. RECENT DEVELOPMENT IN GRAM-NEGATIVE SHOCK

5.1. Endorphins

Faden and Holaday have found that endogenous opiate substances including β-endorphin, a pituitary hormone, may produce significant hypotension and have a pathophysiologic role in the GNS syndrome. They have performed a series of experiments demonstrating that survival of rats and dogs in experimental endotoxin shock is greatly enhanced by the blocking of opiate receptor sites with naloxone (Faden and Holaday 1980).

The pathophysiologic and therapeutic implications of their work are considerable and are discussed in Volume 2, Chapter 12. Sporadic case reports of transient benefit of treatment of varying clinical shock states with naloxone have appeared and are encouraging. Controlled clinical trials of the efficacy of opiate antagonist therapy along with an endorphin assay will be needed before any conclusions can be drawn.

5.2. Endotoxin antibodies

As noted earlier in this chapter (Table 1), the survival of patients with GNS remains dismal and mortality is typically 60%. Ziegler et al have conducted a remarkable clinical study of the efficacy of human endotoxin antibody and have shown a favorable effect on mortality in GNS (Ziegler et al 1982). The study was multicentered and controlled. Patients suspected of being septic were given either human anti-endotoxin sera or preimmune sera (control) from the same human sera donor pool. When the data were analyzed and appropriate exclusions made, endotoxin antibody was shown to be effective in reducing mortality, particularly in patients with GNS. The mortality in the control group of patients with GNS receiving preimmune sera was 76%. The mortality in patients with GNS receiving anti-endotoxin sera was 46%.

The implication is that endotoxin is most likely a major mediator of mortality of GNS in humans and that its harmful effects can be blocked by anti-endotoxin sera in some cases, even when the antisera are given after the endotoxemia has been established.

The efficacy of endotoxin antisera may prove to be the most significant advance in the treatment of GNS in the past 30 years. Confirmation of this important work is needed.

6. SUMMARY

GNS is largely an iatrogenic, nosocomial problem. Alterations in host defense resulting from modern medical therapeutic modalities produce the milieu in which gram-negative sepsis can develop. Despite our awareness of the problem and the availability of effective antibiotics, the mortality of GNS continues to be approximately 60%.

The early hemodynamic response to sepsis is characterized by a decreased SVR with a compensatory increase in cardiac output. As the syndrome progresses, myocardial dysfunction ensues. The early decline in SVR has been found to be associated with activation of prekallikrein and complement in man. The vasoactive phenomena resulting from complement activation may alter regional myocardial blood flow producing segmental, reversible, myocardial dysfunction.

Furthermore, the changes in capillary permeability produced by the activation of vasoactive substances may well explain the many diverse physiologic alterations associated with GNS. Changes in pulmonary capillary permeability may account for the respiratory distress syndrome often noted with GNS. Alterations in capillary integrity may account for the generalized interstitial edema and physiologic arteriovenous shunting accompanying GNS. Disseminated intravascular coagulation may also, in part, be secondary to endothelial damage induced by vasoactive substances.

The experimental demonstration of the ability of naloxone to reduce the mortality in experimental endotoxin shock has opened a new avenue for clinical study.

The demonstration of the efficacy of human anti-endotoxin sera in reducing the mortality of GNS in man provides an important insight into the pathophysiologic mechanisms of GNS and may provide the most important, single therapeutic advance in the past 20 years.

REFERENCES

Blain CM, Anderson TO, Peitras RJ, Gunnar RM (1970) Immediate hemodynamic effects of gram-negative vs. gram-positive bacteremia in man. *Arch. Intern. Med. 126*, 260-265.

Blaisdell FW, Schlobohm RM (1973) The respiratory distress syndrome: a review. *Ann. Surg. 74*, 251-262.

Blaisdell FW, Lim RC, Stallone RJ (1970) The mechanism of pulmonary damage following traumatic shock. *Surg. Gynecol. Obstet. 130*, 15-22.

Bokisch VA, Top FH Jr, Russell PK, Dixon FJ, Muller-Eberhard HJ (1973) The potential pathogenic role of complement in dengue hemorrhage shock syndrome. *N. Engl. J. Med. 289*, 996-1000.

DiScipio RG (1982) The activation of the aternative pathway C3 convertase by human plasma kallikrein. *Immunology 45*, 587–595.

DuPont HL, Spink WW (1969) Infections due to gram-negative organisms: an analysis of 860 patients with bacteremia at the University of Minnesota Medical Center, 1958-1966. *Medicine 48*, 307-332.

Ellrodt AG, Kimchi A, Berman D, Reidinger M, Murata GH (1983) Segmental dysfunction in septic shock (Abstract). *Clin. Res. 31*, 7A.

Faden AI, Holaday JW (1980) Experimental endotoxin shock: the pathophysiologic function of endorphins and treatment with opiate antagonists. *J. Infect. Dis. 142*, 229-238.

Fearon DT, Ruddy S, Schur PH, McCabe WR (1975) Activation of the properdin pathway of complement in patients with gram-negative bacteremia. *N. Eng. J. Med. 292*, 937–951.

Fong JSC, Grod RA (1971) Prevention of the localized and generalized Schwartman reactions by an anti-complementary agent, cobra venom factor. *J. Exp. Med. 134*, 642–655.

Gilbert RP (1962) Endotoxin shock in the primate. *Proc. Soc. Exp. Biol. Med. 111*, 328–331.

Gunnar RM, Loeb HS, Winslow EJ, Blain C, Robinson J (1973) Hemodynamic measurements in bacteremia and septic shock in man. *J. Infect. Dis. 128*, S295-S298.

Harpel PC, Cooper NR (1975) Studies on human plasma Cl inactivater enzyme interactions I. Mechanisms of interaction with Cls, plasmin and trypsin. *J. Clin. Invest. 55*, 593–604.

Hess ML, Krause SM (1981) Contractile protein dysfunction as a determinant of depressed cardiac contractility during endotoxin shock. *J. Mol. Cell. Cardiol. 13*, 715-723.

Hess ML, Hastillo A, Greenfield LJ (1981) Spectrum of cardiovascular function during gram-negative sepsis. *Prog. Cardiovasc. Dis. 23*, 279-298.

Hinshaw LB, Kuida H, Gilbert RP, Visscher MB (1957) Influence of perfusate characteristics on pulmonary vascular response to endotoxin. *Am. J. Physiol. 191*, 293-295.

Hosea S, Brown E, Hammer C, Frank M (1980) Role of complement activation in a model of adult respiratory distress syndrome. *J. Clin. Invest. 66*, 375-382.

Jacobs HS (1981) The role of Activated Complement and granulocytes in shock states and myocardial infarction. *J. Lab. Clin. Med. 98*, 645-653.

Joiner DA, Hawiger A, Gelfand JA (1983) A study of optimal conditions for an assay of the human complement pathway. *Am. J. Clin. Pathol. 79*, 65–72.

Kane MA, May JE, Frank MM (1973) Interaction of the classical and alternative complement pathway with endotoxin lipopolysaccharide. Effect on platelets and blood coagulation. *J. Clin. Invest. 52*, 328–331.

Kimchi A, Ellrodt AG, Berman D, Reidinger M, Murata GH (1983) Right ventricular performance in septic shock (Abstract). *Clin. Res. 31*, 11A.

Kondo Y, Culpepper RD, Hardy JD, Turner MD (1975) Heart transplantation in endotoxin shock. *Surg. Gynecol. Obstet. 140*, 252-254.

Kreger BE, Craven DE, Carling PC, McCabe WR (1980) Gram-negative bacteremia, III. Reassessment of etiology, epidemiology and ecology in 612 patients. *Am. J. Med. 68*, 332-343.

Kwaan HM, Weil MH (1969) Differences in the mechanism of shock caused by bacterial infections. *Surg. Gynecol. Obstet. 128*, 37-45.

Loeb HA, Cruz A, Cheng YT, Boswell J, Pietras RJ, Tobin JR, Gunnar RM (1967) Hemodynamic studies in shock associated with infection. *Br. Heart J. 29*, 883-894.

MacLean LD, Mulligan WG, McLean APH, Duff JH (1967) Patterns of septic shock in man – a detailed study of 56 patients. *Ann. Surg. 166*, 543-562.

Mason DT, Melmon KL (1965) Effects of bradykinin on forearm venous tone and vascular resistance in man. *Circ. Res. 17*, 106-113.

McCabe WR (1974) Gram-negative bacteremia. *Adv. Intern. Med. 19*, 135-158.

McCabe WR, Jackson GG (1962) Gram-negative bacteremia. I. Etiology and ecology. *Arch. Intern. Med. 110*, 847-855.

Nies AS, Forsyth RP, Williams HE, Melmon KL (1968) Contribution of kinins to endotoxin shock in unanesthetized rhesus monkeys. *Circ. Res. 22*, 155-164.

Nishijima H, Weil MH, Shubin H, Cavanilles J (1973) Hemodynamic and metabolic studies on shock associated with gram-negative bacteremia. *Medicine 52*, 287-294.

Osterud B, Bjorklid E (1982) Human factor VII associated with endotoxin stimulated monocytes in whole blood. *Biochem. Biophys. Res. Commun. 108*, 620–626.

Robinson JA, Klodnycky ML, Loeb HS, Racic MR, Gunnar RM (1975) Endotoxin prekallikrein, complement and systemic vascular resistance. *Am. J. Med. 59*, 61-67.

Scott WW (1929) Blood stream infection in urology: a report of 82 cases. *J. Urol. 21*, 527-566.

Siegel JH, Greenspan M, Del Guercio LRM (1967) Abnormal vascular tone, defective oxygen transport and myocardial failure in human septic shock. *Ann. Surg. 165*, 504-517.

Sullivan NM, Sutter VL, Mims MM, Marsh VH, Finegold SM (1973) Clinical aspects of bacteremia after manipulation of the genitourinary tract. *J. Infect. Dis. 127*, 49-55.

Udhoji VN, Weil MH (1965) Hemodynamic and metabolic studies on shock associated with bacteremia. Observations on 16 patients. *Ann. Intern. Med. 62*, 966-978.

Vito L, Dennis RC, Weisel RD, Hechtman HB (1974) Sepsis presenting as acute respiratory insufficiency. *Surg. Gynecol. Obstet. 138*, 896-900.

Waisbren BA (1951) Bacteremia due to gram-negative bacilli other than the salmonella. A clinical and therapeutic study. *Arch. Intern. Med. 88*, 467-488.

Weisel RD, Vito L, Dennis RC, Valeri CR, Hechtman HB (1977) Myocardial depression during sepsis. *Am. J. Surg. 133*, 512-521.

Winslow EJ, Loeb HS, Rahimtoola SH, Savitri K, Gunnar RM (1973) Hemodynamic studies and results of therapy in 50 patients with bacteremic shock. *Am. J. Med. 54*, 421-432.

Wolff SM, Bennett JV (1974) Gram-negative rod bacteremia. *N. Engl. J. Med. 291*, 733-734.

Ziegler EJ, McCutchan JA, Fierer J, Glauser MP, Sadoff JC, Douglas H, Braude AI (1982) Treatment of gram-negative bacteremia and shock with human antiserum to a mutant Escherichia coli. *N. Engl. J. Med. 307*, 1225-1230.

Handbook of Endotoxin, Vol. 4: Clinical Aspects of Endotoxin Shock
R.A. Proctor, editor
© Elsevier Science Publishers B.V., 1986

CHAPTER 2

Pathologic manifestations of septic shock

LINDA T. ARCHER

1. DEFINITIONS OF SHOCK

Maclean et al (1965) expressed uniquely that shock is a failure of blood flow and not of blood pressure. Visscher (1972) underscored this view in a symposium introduction entitled 'An Overview of the Shock Problem', stating that underperfusion of some tissue beds is a common denominator in all forms of circulatory shock and in all species of animals. Robbins and Cotran (1979a) define shock as a profound hemodynamic and metabolic disturbance produced by any massive insult to the body, such as profuse hemorrhage, severe trauma or burns, extensive myocardial infarction, massive pulmonary embolism, and uncontrolled bacterial sepsis, to mention only the most important precursors. In reality, it is a constellation of syndromes, all characterized by low-perfusion circulatory insufficiency leading to an imbalance between the metabolic needs of vital organs and the available blood flow (Robbins and Cotran 1979a). As a result, oxygen and nutrient delivery to cells and removal of waste products are decreased. Depending on the severity of these deficits and their duration, the cells of many vital organs suffer injury and even death.

Clinically, shock is classified on the basis of the precipitating event, for example, 'traumatic shock', 'burn shock', 'septic shock', or 'cardiac shock'. 'Septic shock' is further classified as either overwhelming gram-positive infections or overwhelming gram-negative infections with endotoxemia (endotoxin shock) (Robbins and Cotran 1979a).

2. HISTORY OF ANIMAL MODELS: ENDOTOXIN VERSUS LIVE ORGANISMS

Spink et al (1948) first indicated that endotoxin participates in the precipitation of bacteremic shock: they observed fever, systemic hypotension, and

18

tachycardia as a consequence of administering antibiotics to bacteremic patients. Their work stimulated others to use endotoxin to precipitate shock in experimental animal studies, and the subsequent reports contributed markedly to the understanding of many of the mechanisms involved in the pathogenesis of septic shock (Gilbert 1962; Kuida et al 1961; Maclean and Weil 1956; Maclean et al 1956; Weil et al 1956). Waisbren (1964) pointed out some differences between the responses of man and dogs to endotoxin. He concluded that: 'In the last analysis, the true explanation of the pathophysiology of gram-negative shock will come from more intensive studies of patients while they are suffering from this condition and perhaps more practically, from intensive studies of an animal model that more closely simulates the clinical condition in man (...). These further studies (...) may prove that endotoxin has a role in gram-negative shock, but (...) the equation of endotoxin and gram-negative shock seems unwarranted.'

Over the years, it has become evident that septic shock is too complicated a process to limit its study *only* to humans with the disease: 'There are formidable limitations in studying bacterial shock in patients rather than in experimental animals, for example, the (human) subjects are sick persons who have had infections for a variable period of time and previous treatment is a factor which is entirely uncontrolled' (Udhoji et al 1963). Moreover, the study of human patients is further complicated by variables such as: differences in age; nutritional or immunological status; preexisting problems such as cancer or pulmonary, hepatic, or renal disease; or prior insults such as trauma, burns, or surgery (Hinshaw et al 1985).

Other investigators have also voiced significant concerns regarding the relevance of producing 'septic shock' by injecting animals with endotoxin only (Hinshaw 1972; Morrison and Ulevitch 1978; Powrie and Norman 1976; Rietschel et al 1982; Sibbald 1976). In general, endotoxin is still considered the primary factor that precipitates the shock state (Cavanagh et al 1977), but other mechanisms may play fundamental or contributory roles (Spink 1972). It has been suggested that live organisms and endotoxin invading the bloodstream in combination may elicit a Shwartzman phenomenon or an acute allergic reaction, either of which may lead to shock (Waisbren 1964).

3. HEMODYNAMIC ALTERATIONS

3.1. Overview

The hemodynamic responses to gram-negative organisms or endotoxin are quite complex and result in many physiological alterations. Inadequate perfusion of vital organs results from these alterations and prevents cells from maintaining a steady metabolic status. When this happens, multiple organ systems can fail to function (Cerra et al 1979b, 1982; Hassett et al 1983) and death can occur.

Four sequential phases of cardiovascular system involvement have been documented in animal septic shock models that correlate with those observed in human patients (Hinshaw 1982). First, the pre-shock period is characterized by a moderate decrease in mean systemic arterial pressure, normal or increased cardiac output, tachycardia, tachypnea, fever, restlessness, and irritation. The second phase is marked by progressively developing systemic hypotension and decreased cardiac output. A compensatory period, triggered by prolonged systemic hypotension, follows, and the heart often demonstrates dysfunction. The final phase can be either eventual improvement of metabolic-endocrinologic, host defense, and hemodynamic functions resulting in survival, or progressive circulatory deterioration resulting in multiple organ failure and death.

3.2. Myocardial dysfunction

Clinical and experimental data concur that myocardial dysfunction occurs during the intermediate and latter phases of septic shock (Archer et al 1982; Guntheroth et al 1982; Hinshaw et al 1972b, 1979b; Siegel et al 1967; Weisel et al 1977; Weisul et al 1975). Inadequate coronary perfusion may be a cause of the dysfunction, and endotoxin can act synergistically with low coronary perfusion pressure to exacerbate the degree of dysfunction (Hinshaw et al 1974a). Other factors associated with dysfunction include edema formation in the contractile elements and/or mitochondria, ionic imbalances of potassium and calcium, and depressed myocardial responses to β-adrenergic stimuli (Archer et al 1975, 1978a; Coalson et al 1972; Hinshaw et al 1979b). Endotoxin itself or toxic humoral byproducts may also depress myocardial function (Hess and Krause 1979; Wilson 1980).

3.3. Classification of septic shock on the basis of hemodynamic criteria

Weil (1972, 1977) has developed the concept that there are four general forms of shock: hypovolemic, cardiogenic, distributive, and obstructive. Each of these classifications has a 'hemodynamic' basis. He places gram-negative bacterial or endotoxin shock in the category of 'distributive shock' in that it results from a maldistribution of blood volume. According to this view, blood is redistributed within the vascular bed during septic shock resulting in lowered cardiac output and organ malperfusion. Weil states that his classification must be regarded as provisional and may change if the most current hypotheses and extrapolations from animal experimentation are not corroborated. This distributive form of shock can be divided into: (a) low vascular resistance or (b) high or normal vascular resistance patterns. Both patterns usually demonstrate perfusion: ventilation defects, and in some instances, myocardial depression. The low-resistance form is usually associated with normal or increased cardiac output during the early ('warm shock') phase, and the high-resistance form ('cold phase') with lowered cardiac output, decreased blood volume, and elevated pulmonary and systemic vascular resistances (Cerra et al 1979a; Clowes et al 1974; MacLean et al 1967; Nishijima et al 1973; Siegel et al 1971, 1979; Udhoji and Weil 1965; Weil 1972, 1977).

In patients with the normal- or high-resistance type of shock in which cardiac output is reduced, the 'distributive defect' is sequestration of blood in the venous circuit (Weil 1977). In low-resistance type shock in which cardiac output is increased, the 'distributive defect' includes vasodilatation and arteriovenous shunting (Weil 1977). The high-output type is more often observed in the early febrile stages of bacillary bacteremia prior to onset of clinical shock.

These hemodynamic phases, 'hyperdynamic', low resistance/high output, or 'hypodynamic', high resistance/low output, have been emphasized in the clinical literature because the 'stage', 'state', or 'phase' of shock observed in a septic patient is extremely critical in terms of survival rate. The hyperdynamic (high-flow) and hypodynamic (low-flow) states represent, in general, the difference between survival (the former state) and nonsurvival (the latter state) since increased mortality and decreased cardiac output are clearly related (Cerra et al 1979a; Clowes et al 1974; MacLean et al 1967; Nishijima et al 1973; Siegel 1971, 1979; Weil 1972, 1977). In fact, Nishijima et al (1973) summarized the results of seven clinical reports and showed that as cardiac output decreased, mortality rate increased. If therapeutic intervention fails during the hyperdynamic phase or if intrinsic protective physiological mechanisms are insufficient, then the hyperdynamic phase can deteriorate into the hypodynamic phase (Hinshaw 1982). The hypodynamic phase can also occur without progressing through the hyperdynamic phase if the septic insult is overwhelming (Hinshaw 1982). Once in the hypodynamic phase, a progression of pathophysiological positive feedback pathways results in multiple 'vicious cycles' that usually culminate in multiple organ failure and death (Cerra et al 1979b, 1982; Hassett et al 1983; Hinshaw 1982; Wilson 1980). During the hypodynamic phase, hypovolemia results from maldistribution of blood and exacerbates the existing physiological disorders (Wilson 1980). Progressive increase of venous capacitance causes the circulating blood volume to decrease, thereby further exacerbating the 'distributive defect' (Weil 1972, 1977).

4. ANIMAL SPECIES USED IN SEPTIC SHOCK MODELS

4.1. Clinical correlates of dog and baboon models

Two animal species that have proved to be particularly valuable in septic shock research are the dog and baboon. Investigators have documented that the dog and baboon respond to septic shock much like man in that they all exhibit varying degrees of hypotension (Archer et al 1978b; Balis et al 1974, 1978; Brobmann et al 1970; Coalson et al 1975; Hinshaw et al 1968, 1970a, 1970b, 1974c, 1977, 1978, 1979a, 1980, 1981a, 1981b, 1983; MacLean and Weil 1956; MacLean et al 1956, 1967; Richman et al 1980; Siegel et al 1967; Udhoji and Weil 1965; Udhoji et al 1963; Waisbren 1964; Weil 1972, 1977; Weil et al 1956), thrombocytopenia (Coalson et al 1979; Hinshaw et al 1980, 1981a, 1983; Milligan et al 1974), leukopenia or granulocytopenia (Archer et al 1978b; Balis et al 1974, 1978; Coalson et al 1978b; Hinshaw et al 1977, 1980, 1981a, 1981b, 1983; Kreger et al 1980; Milligan et al 1974; Waisbren 1964; Wilson 1976; Weil 1977), and abnormalities of carbohydrate metabolism (Archer et al 1978b; Cerra et al 1979b, 1982; Cryer et al 1971, 1972; Clowes et al 1974, 1978; Groves et al 1974; Hinshaw et al 1974c, 1977, 1978, 1979a, 1980, 1981a, 1981b). In addition, the baboon has been reported to respond like man to septic shock in that Hageman factor can be activated (Mason and Colman 1971; Nies et al 1968; Robinson et al 1975) leading to coagulative changes (Morrison and Ulevitch 1978) and disseminated intravascular coagulation (DIC) (Balis et al 1974; 1978; Coalson et al 1975, 1979; Corrigan et al 1968; Hardaway 1978; McKay et al 1967; Milligan et al 1974; McGovern and Tiller 1980; Remmele and Harms 1968; Richman et al 1980; Robboy et al 1972). One of the most significant parallels is the discovery that the baboon develops an early 'hyperdynamic phase' during lethal *Escherichia coli* infusion (Hinshaw et al 1983). Within 2 hours, increases in cardiac output, oxygen uptake, heart rate, rectal temperature, respiration rate, and respiration minute volume occur ($p < 0.05$), while mean arterial pressure, total peripheral resistance, pulmonary artery and pulmonary artery wedge pressures, Pa_{CO_2} and arterial and venous CO_2 contents decrease ($p < 0.05$). The human patient in sepsis or septic shock often exhibits an early 'hyperdynamic phase' characterized by decreased peripheral vascular resistance, increased cardiac output and tachycardia (Clowes et al 1974, 1978; MacLean et al 1967; Siegel et al 1967, 1971; Udhoji et al 1963; Udhoji and Weil 1965; Weil 1972, 1977; Wilson 1976). The one apparent dissimilarity in response between dog and man to septic shock is the marked hepatosplanchnic pooling and eventual massive hemorrhagic necrosis that occurs in the gastrointestinal tract of the dog (Archer et al 1978b; Brobmann et al 1970; Hinshaw 1968; Hinshaw et al 1974c, 1977, 1979a; Kuida et al 1961; Marston 1962; McGovern and Tiller 1980; Reynolds et al 1972).

4.2. Therapy for lethal septic shock in the dog and baboon

Hinshaw et al (1979a, 1980) have documented that intermittent infusions of methylprednisolone sodium succinate (MPSS) and gentamicin sulfate (GS) result in 100% survival of dogs and baboons after challenge with LD_{100} *E. coli*, provided the treatment is initiated soon after the septic insult. Sequential delay of initiation of MPSS administration from 30 to 120 to 240 minutes after the onset of a 2-hour *E. coli* infusion in the baboon results in 100%, 85%, and 65% survival, respectively (Hinshaw et al 1980, 1981a, 1981b). MPSS/GS therapy also results in 100% survival in the dog infused for 1 hour with LD_{100} *E. coli* if MPSS infusion is initiated 15 minutes after the onset of *E. coli* infusion (Hinshaw et al 1979a). Neither the dog (Hinshaw et al 1979a) nor the baboon (Coalson et al 1978b; Hinshaw et al 1978, 1980) infused with LD_{100} *E. coli* has been prevented from dying when treated with GS therapy only (except in rare instances) or with MPSS therapy only.

4.3. Morphological evaluation in the baboon

In a subsequent study, tissues taken at autopsy from the baboons in the three sequential MPSS/GS treatment studies including untreated baboons and those treated with GS only were evaluated (Archer et al 1983). The mor-

TABLE 1 *Distribution of morphologic lesions in LD_{100} E. coli-shocked baboons (N = 27)*

Site of lesion	Nonsurvivors (N = 14)		Survivors (N = 13)	
	Animals with lesions (%)	Scores for lesions*	Animals with lesions (%)	Scores for lesions*
Adrenal				
cortical: congestion	100	1-4	54	1-2
hemorrhage	100	1-4	0	-
medullary: congestion	50	1-4	8	1
hemorrhage	40	1-4	0	-
Liver				
increased PMN	100	2-3	30	1
congestion	93	2-4	30	1
fibrin thrombi	20	-	0	-
Kidney				
cortical tubular necrosis	64	1-3	15	1
congestion	42	2-4	0	-
hemorrhage	21	1-3	0	-
increased PMN	21	1-3	30	1-2
fibrin thrombi	30	-	0	-
Lung				
increased PMN	57	2-3	8	1
congestion	64	2-3	54	1-2
edema	57	1-4	8	2
hemorrhage	50	1-4	0	-
fibrin thrombi	36	-	8	-

* +1: mild, +2: moderate, +3: severe or marked, and +4: massive.
PMN = polymorphonuclear leukocytes.

phologic manifestations of shock (seen in nonsurvivors whether treated or not) indicate inflammatory response, maldistribution of blood, disseminated intravascular coagulation, and multiple organ damage. The tissue lesions observed in nonsurviving baboons included combinations of mild to massive congestion, edema, hemorrhage, fibrin thrombi, increased numbers of polymorphonuclear leukocytes (PMN) and necrosis. The main organs injured by the *E. coli*-induced shock were adrenal glands, liver, kidneys, lungs, and occasionally, the spleen (see Table 1 for details). However, when the treatment was initiated during *E. coli* infusion, the multiple organ damage was prevented or reduced in 100% of the animals. When initiation of treatment was delayed until all *E. coli* was infused or 2 hours after that, the lesions were prevented or reduced in 85% and 65% of the baboons respectively. The lesions of the nonsurviving treated baboons were indistinguishable from those of the untreated baboons, none of whom survived.

5. PERFUSION DEFICITS CONTRIBUTING TO TISSUE LESIONS

Regardless of the animal species used or the substance used to induce septic shock, the result is inadequately perfused tissues. In fact, in some cases, circulation comes almost to a standstill (Hinshaw 1972; Hinshaw et al 1968, 1980, 1981a, 1981b; MacLean and Weil 1956; Siegel 1967; Weil 1972, 1977; Weil et al 1956; Wilson 1980). The hypoxic tissues convert to anaerobic metabolism and because of the alteration in hemodynamics, products of metabolism accumulate. Because blood flow rate decreases through the affected organs, there may be a lag time before the accumulating metabolites begin to appear in the peripheral circulation (Bergentz et al 1969), but eventually blood lactate concentrations increase (Cerra et al 1979b; Clowes et al 1974; Hinshaw et al 1980, 1981a, 1981b; Udhoji and Weil 1965; Wilson 1980). The combination of tissue anoxia and acidosis is very harmful to cell organelles and may lead to irreversible tissue damage (Mela et al 1971; Trump 1975). Additionally, in the presence of hypoxia and acidosis, tissue lysosomal membranes disrupt and lysosomal enzymes are released causing further tissue damage (Janoff 1964; Slater 1969).

In addition, when blood flow through an organ is compromised, the accumulation of mediators intensifies the shock process (Altura and Halevy 1977; Hinshaw 1971, 1982; Morrison and Ulevitch 1978; Thal 1972). Platelets and neutrophils are more likely to marginate and aggregate and release vasoactive or toxic substances, that is, lysosomal enzymes (Cohn and Hirsch 1960), leukotrienes (Jubiz et al 1982), O_2 radicals (Goldstein et al 1975), thromboxane (Hamberg et al 1975), and serotonin (Kobold et al 1964). Monocytes/macrophages can also release mediators (Unanue 1976), for example, procoagulant substances (Rivers et al 1975; Hiller et al 1977), lysosomal enzymes (Wahl et al 1974), plasminogen activator (Gordon et al 1974), pyrogens (Dinarello et al 1974), and prostaglandins (Kurland and Bockman 1978). These mediators can directly cause tissue injury or can intensify the flow deficit and cause further tissue damage through hypoxia/acidosis.

6. TISSUE LESIONS

The histological manifestations of shock include congestion, edema, fibrin thrombi, hemorrhage, and necrosis. Congestion is an excessive or abnormal accumulation of blood in the lumen of dilated vessels. In septic shock, blood is selectively sequestered in the venous circuit (Weil 1972, 1977), which may account for the congestion observed in the various organs. Edema is a condition in which abnormally large amounts of fluid accumulate in intercellular tissue spaces of the body. Edema results from an imbalance between forces that tend to retain fluid in the intravascular compartment and those that tend to retain fluid in the interstitial tissue spaces, for example, hydrostatic pressure in the capillary and oncotic pressure in interstitial fluid opposed by hydrostatic pressure in the interstitial fluid and oncotic pressure due to plasma proteins (Berne and Levy 1977; Robbins and Cotran 1979a). Edema also results from increased capillary permeability (Berne and Levy 1977). Fibrin thrombi are aggregations of primarily fibrin attached to the walls of blood vessels that frequently cause partial vascular obstruction at the point of their formation. Fibrin thrombi are the hallmarks of disseminated intravascular coagulation (DIC). The phenomenon of DIC, although associated with all types of shock, is most prominent in septic shock (Archer et al 1983; Balis et al 1974, 1978; Coalson et al 1975, 1979; Corrigan et al 1968; Hardaway 1978; McGovern and Tiller 1980; McKay et al 1967; Milligan et al 1974; Remmele and Harms 1968; Robbins and Cotran 1979a; Robboy et al 1972). Hemorrhage is the presence of blood outside the vessels. Necrosis is death of tissue, usually as individual cells, groups of cells, or in small localized areas. In septic shock, the necrosis is often focal and is often the result of ischemia or presence of organisms (McGovern and Tiller 1980; Moon 1948; Robbins and Cotran 1979a; Sandritter and Lasch 1967).

7. SPECIFIC ORGAN DISTRIBUTION OF LESIONS: A COMPARISON OF MAN*, BABOON **, AND DOG***

The morphologic lesions and the pattern of organ damage observed in non-human primates that die from gram-negative shock (Archer et al 1983; Balis et al 1974, 1978; Coalson et al 1975, 1978b, 1979; Hinshaw et al 1980; McKay et al 1967; Richman 1980) are very similar to those found in tissues removed from humans that die of septic shock, (Blaisdell 1974; Bohm 1982; Dalgaard 1960; Friderichsen 1918; Greendyke 1965; Harms and Lehmann 1969; Lindqvist et al 1963; Marston 1962; Martland 1944; Masland and Barrows 1962; McKay et al 1959; McGovern 1971; McGovern and Tiller 1980; Moon 1948; Orell 1971; Remmele and Harms 1968; Rich 1944; Riede et al 1978, 1981; Robbins and Cotran 1979a; Robboy et al 1972; Sandritter and Lasch 1967; Vassalli and Richet 1960; Waterhouse 1911; Xarli 1978). The morphologic lesions commonly observed in both man and nonhuman primates include congestion, edema, fibrin thrombi, hemorrhage and necrosis in liver, lungs, kidneys, adrenal glands, and spleen (McGovern and Tiller 1980; Sandritter and Lasch 1967). The dog responds like man and baboon in many ways to endotoxin- or live organism-induced shock with one distinct exception: marked hepatosplanchnic pooling and eventual massive hemorrhagic necrosis of the gastrointestinal tract (Archer et al 1978b; Brobmann et al 1970; Brunson et al 1966; Hinshaw 1968; Hinshaw et al 1970b, 1974b, 1977, 1979a; Kuida et al 1961; Lillehei and MacLean 1959; MacLean et al 1956; Reynolds et al 1972).

7.1. Adrenal glands

Fibrin thrombi, hemorrhage and/or necrosis have been repeatedly observed in the adrenal cortex of humans that die of septic shock (Bohm 1982; Friderichsen 1918; Greendyke 1965; Lindqvist et al 1963; Martland 1944;

---

*   Based on literature review
**   Based on an extensive light microscopic evaluation
***   Based on literature review

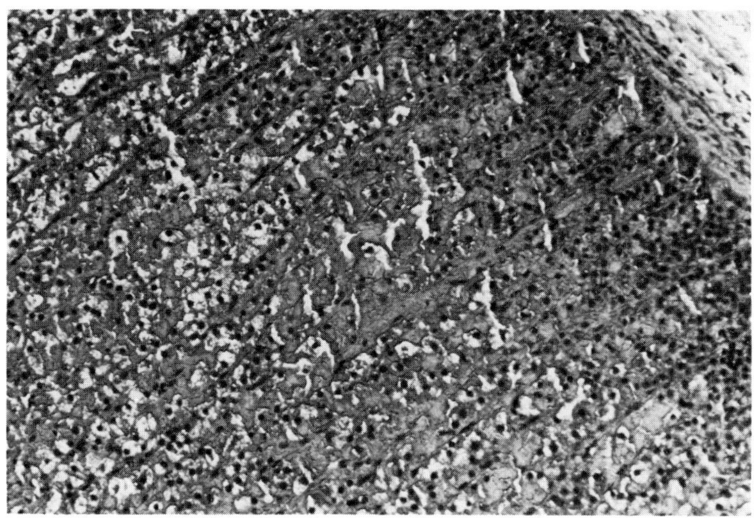

Fig. 1. *Adrenal gland representative of nonsurvivors in all groups. Hemorrhage is present in the cortex, specifically in the zonae glomerulosa and fasciculata. The architecture of the underlying cortical cells is essentially obliterated by the hemorrhage. Hematoxylin and eosin, × 48. Reproduced from Archer et al (1983), with permission.*

McKay et al 1959; McGovern 1971; McGovern and Tiller 1980; Moon 1948; Rich 1944; Sandritter and Lasch 1967; Waterhouse 1911; Xarli 1978). McGovern (1971) states that one of the two most common sites for hemorrhagic lesions resulting from shock in humans is the adrenal gland (the other being the lung). Archer et al (1983) found the morphologic hallmark of nonsurvival in the baboon septic shock model is hemorrhage in the adrenal cortex (Fig. 1).

It has been reported by Brunson et al (1966) that the adrenal glands of dogs injected intravenously with LD_{80} endotoxin begin to show hemorrhage within 5–12 hours and the severity of the hemorrhage increases during the next 12 hours. The hemorrhagic lesion of the adrenal gland involves the outer medulla and inner cortex. In contrast, Kosanke et al (1982) examining the adrenal glands of dogs given LD_{80} live *E. coli* organisms, rarely found hemorrhagic lesions.

28

7.2. Liver

Liver damage has been documented in humans that die from shock (Harms and Lehmann 1969; Lindqvist et al 1963; Masland and Barrows 1962; McGovern 1971; McGovern and Tiller 1980; Moon 1948; Remmele and Harms 1968; Robbins and Cotran 1979a; Sandritter and Lasch 1967). Remmele and Harms (1968) found fibrin thrombi and increased numbers of PMN in the liver sinusoids, and McGovern and Tiller (1980) reported zonal necrosis of the liver in patients dying with septic shock. Archer et al (1983) reported that the lesions in livers of nonsurviving baboons challenged with *E. coli* are increased numbers of PMN, congestion (Fig. 2) and fibrin thrombi (Fig. 3). Coalson et al (1975, 1978a, 1979) found fibrin thrombi and increased numbers of PMN in the liver sinusoids of baboons subjected to endotoxin and live *E. coli*-induced shock. Balis et al (1974, 1978) have confirmed these results in the baboon subjected to prolonged endotoxemia.

The earliest detectable lesions in the dog subjected to LD_{80} endotoxin shock include marked congestion of the liver, and edema and hemorrhage in the gallbladder (Brunson et al 1966). Five to twelve hours after endotoxin administration, hepatic and gallbladder lesions increase in severity and are characterized by areas of hepatic necrosis and profound gallbladder edema. Archer et al (1978b) corroborated these findings when evaluating tissues from dogs injected with twice the usual lethal dose of endotoxin. They observed central lobular necrosis of the liver and increased numbers of PMN.

7.3. Kidneys

The kidneys are frequently damaged in patients with severe shock (Dalgaard 1960; Harms and Lehmann 1969; Lindqvist et al 1963; Masland and Barrows 1962; McKay et al 1959; McGovern 1971; McGovern and Tiller 1980; Moon 1948; Orell 1971; Remmele and Harms 1968; Robbins and Cotran 1979a, 1979b; Robboy et al 1972; Sandritter and Lasch 1967; Vassalli and Richet 1960); tubules at all levels of the nephron are affected and the result is referred to as acute tubular necrosis (ATN) (Robbins and Cotran 1979a). Ischemic ATN is characterized by focal tubular necrosis at multiple points along the nephron (Robbins and Cotran 1979b). In association with the necrosis, often basement membranes are ruptured (tubulorrhexis) and tubular lumina are occluded with casts. The straight portion of the proximal

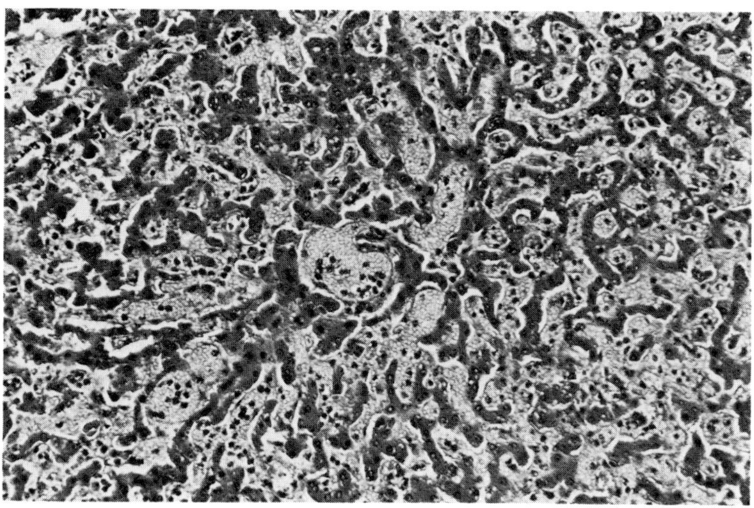

Fig. 2. *Liver from a nonsurvivor. The most consistent liver changes were congestion of and increased numbers of polymorphonuclear leukocytes (PMN) in central veins and sinusoids. Note marked hepatocellular compression atrophy. Hematoxylin and eosin, × 48. Reproduced from Archer et al (1983), with permission.*

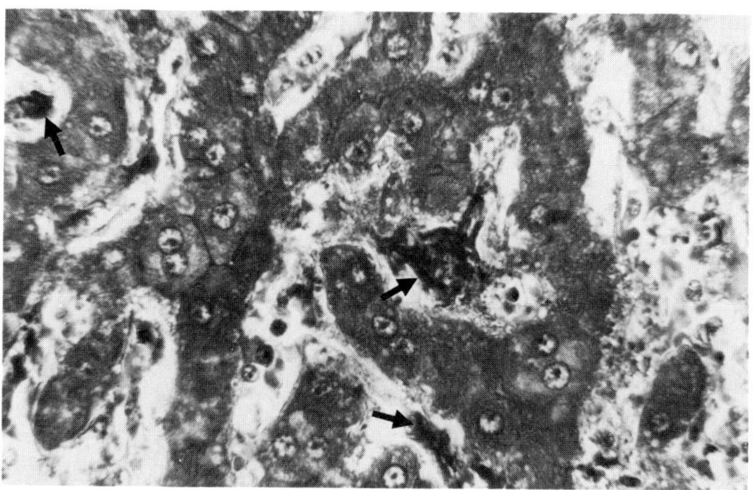

Fig. 3. *Fibrin thrombi (arrows) in ectatic hepatic sinusoids of a nonsurvivor. Phosphotungstic acid-hematoxylin, × 192. Reproduced from Archer et al (1983), with permission.*

tubule is especially vulnerable, but focal necrotic lesions with casts may also occur in the distal nephron.

Eosinophilic hyaline casts, as well as pigmented granular casts, are extremely common, particularly in distal tubules and collecting ducts (Dunnill 1974). These casts consist principally of Tamm-Horsfall protein (Hoyer and Seiler 1979) (a specific urinary glycoprotein normally secreted by cells of the ascending thick limb and distal tubules) and hemoglobin, myoglobin, and other plasma proteins. Other lesions found in ischemic ATN are interstitial edema and accumulations of leukocytes within dilated vasa recta.

Also observed in some cases of human septic shock are fibrin or hyaline thrombi inside the glomerular loops (Lindqvist et al 1963; McKay et al 1959; McGovern 1971; McGovern and Tiller 1980; Moon 1948; Orell 1971; Remmele and Harms 1968; Robbins and Cotran 1979a; Robboy et al 1972; Sandritter and Lasch 1967; Vassalli and Richet 1960), degenerated tubular epithelium, and even total necrosis of tubules (Moon 1948; Sandritter and Lasch 1967).

Archer et al (1983) have shown that lesions observed in the kidneys of baboons that die of septic shock include moderate to severe cortical tubular necrosis (Fig. 4); moderate to massive congestion of cortical, medullary, and glomerular vessels; increased numbers of PMN (+1 to +3*) within vessels and the interstitium; hemorrhage into the interstitium, granular tubular casts, proteinuria (Fig. 5), and/or glomerular fibrin thrombi (Figs. 6 and 7). Coalson et al (1975, 1978b, 1979) observed tubular lesions and glomerular fibrin thrombi in baboons during live *E. coli*-induced shock. Richman et al (1980) have reported that the peritubular capillaries are a major site of endotoxin-induced vascular injury in the primate kidney and that the 'inflammatory' process involving the PMN plays a basic role in the development of acute tubular necrosis. Brunson et al (1966) did not discuss the kidney in their report of endotoxin-shocked dogs.

7.4. Lungs

Detailed histologic descriptions of the lungs of humans that died of septic shock have also been reported (Blaisdell 1974; Harms and Lehmann 1969; McGovern 1971; McGovern and Tiller 1980; Orell 1971; Remmele and Harms 1968; Riede et al 1978, 1981; Robbins and Cotran 1979a; Sandritter

---

* Numerical rating system for lesions used in this chapter: +1 = mild; +2 = moderate; +3 = severe or marked; +4 = massive.

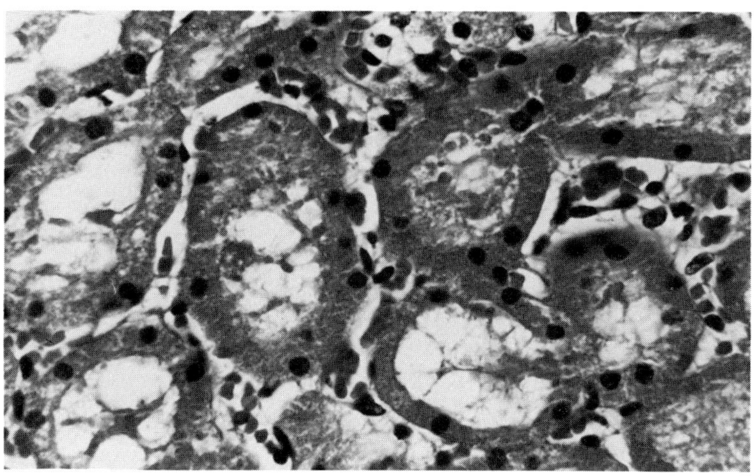

Fig. 4. *The most frequent kidney lesion of nonsurvivors in all groups: cortical tubular necrosis. The proximal convoluted tubular cell cytoplasm is vacuolated, swollen and contains hyaline droplets. The tubular epithelial cell nuclei are pyknotic. Hematoxylin and eosin, × 192. Reproduced from Archer et al (1983), with permission.*

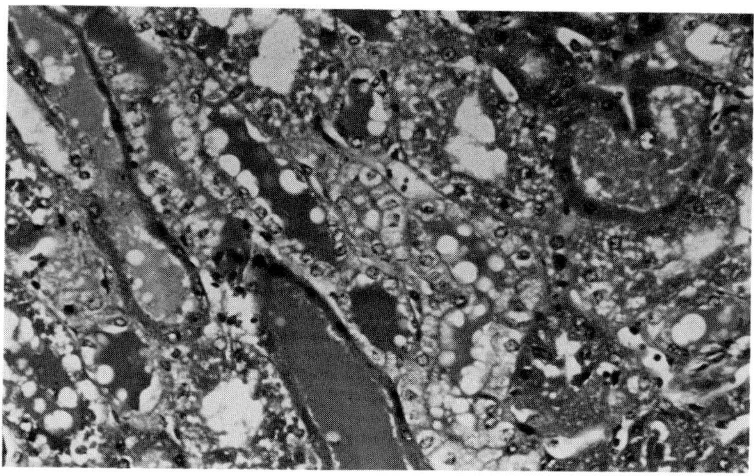

Fig. 5. *Distal segment of kidney tubules of a nonsurvivor. Distal tubular cell cytoplasm is vacuolated, swollen and hyalinized. Nuclei are pyknotic and karyolytic, and numerous dead cells are desquamated into lumina forming granular casts. Note also marked proteinuria. Hematoxylin and eosin, × 72. Reproduced from Archer et al (1983), with permission.*

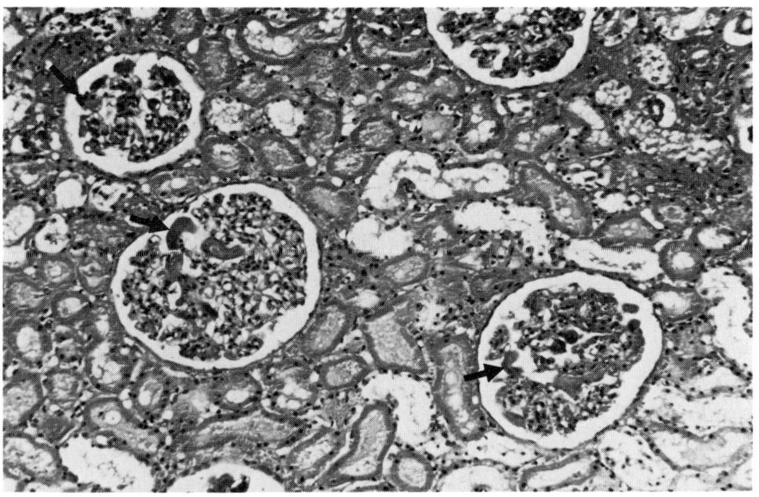

Fig. 6. *Kidney from baboon dying in 13 hours. Dark amorphous masses, consistent with fibrin thrombi, are present in arterioles of many glomeruli (arrows). Tubular cell changes as described in Figures 4 and 5 are also prominent. Hematoxylin and eosin, × 48. Reproduced from Archer et al (1983), with permission.*

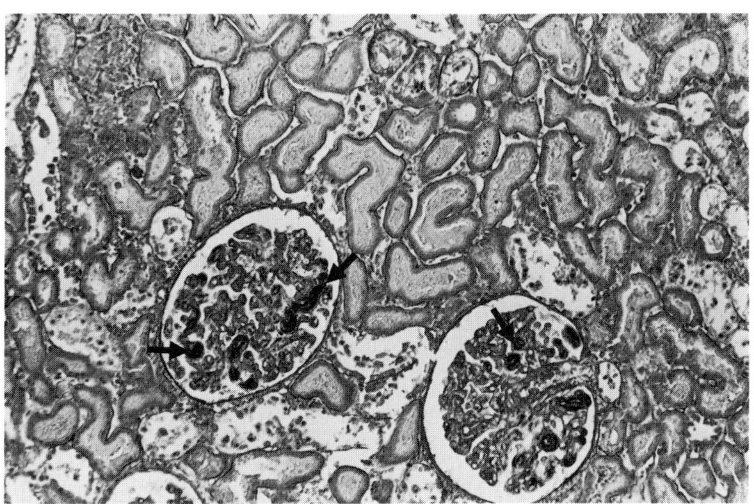

Fig. 7. *Kidney same as in Figure 6. Fibrin thrombi are present in arterioles of glomeruli (arrows). Tubular epithelial cell desquamation and interstitial hemorrhage are also present. Phosphotungstic acid-hematoxylin, × 48. Reproduced from Archer et al (1983), with permission.*

and Lasch 1967). Orell (1971) evaluated lungs of patients who died with respiratory distress associated with shock and trauma and found interstitial and alveolar edema and thickening of the capillary walls due to congestion and diapedesis of red blood cells. He also observed hemorrhage, hyaline membranes and microscopic thrombi in many lungs. Remmele and Harms (1968) and McGovern and Tiller (1980) also have reported fibrin thrombi in the lungs of patients. Blaisdell (1974) performed 200 autopsies on patients who died at various times after the onset of respiratory distress and described the evolution of lung lesions as follows: clumping of white blood cells, platelets, and fibrin in pulmonary arterioles (0–6 hours); severe congestion, interstitial edema, and hemorrhage with intra-alveolar edema and hemorrhage (24 hours); and extensive alveolar edema and hemorrhage (48 hours). Nonsurviving baboons infused with *E. coli* (Archer et al 1983) show a similar progression of lung lesions: increased numbers of PMN (+2) and moderate congestion (5–7 hours) (Fig. 8); PMN (+3), severe congestion, mild to severe edema, moderate hemorrhage and fibrin thrombi (10–15 hours); and massive diffuse alveolar edema and hemorrhage (49 hours) (Fig. 9).

Brunson et al (1966) have reported that pulmonary congestion, edema and hemorrhages begin to develop 5–12 hours after endotoxin challenge in the dog. These lesions increase in severity during the next 12 hours post-endotoxin and become most severe after 24 hours.

7.5. Intestine

McGovern and Tiller (1980) have shown that the intestinal tract is the third most commonly involved system in man. The characteristic lesion is acute ischemic enterocolitis. The appearance of ischemic enterocolitis can vary a great deal. The affected portions of the intestine are intensely congested and edematous and frequently blood is found free in the lumen. Ulcers may be punched out, transverse, linear, occasionally perforated, and sometimes have wide areas of denudation.

The small intestine, if affected, is usually more severely affected than the large intestine (Marston et al 1966). The segments most commonly affected by shock are the splenic flexure and descending colon (Marston 1962; Marston et al 1966).

The microscopic picture of ischemic enterocolitis includes edema and fibrinoid material in and around the walls of venules and veins of the submucosa. Thrombi are occasionally present as well as a dense inflammatory infiltrate consisting of neutrophils in the vessel walls and surrounding submucosa.

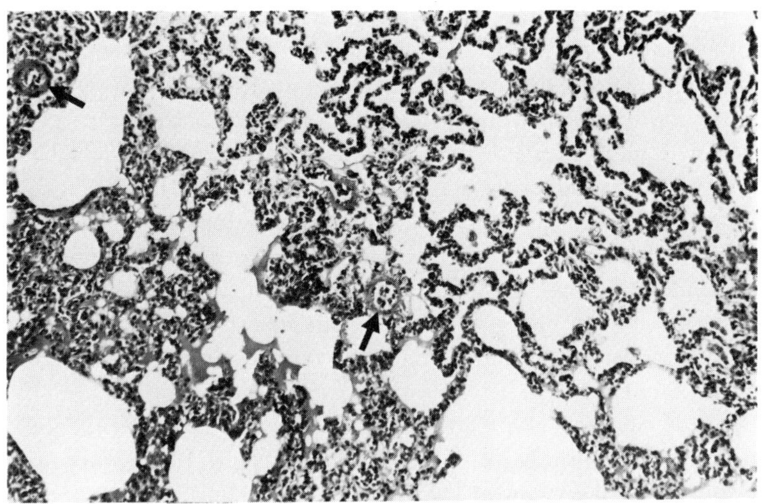

Fig. 8.   *Lung from a nonsurvivor. Alveolar septa are congested and contain increased numbers of PMN. Additionally PMN are marginating along the arterial walls (arrows). Hematoxylin and eosin, × 48. Reproduced from Archer et al (1983), with permission.*

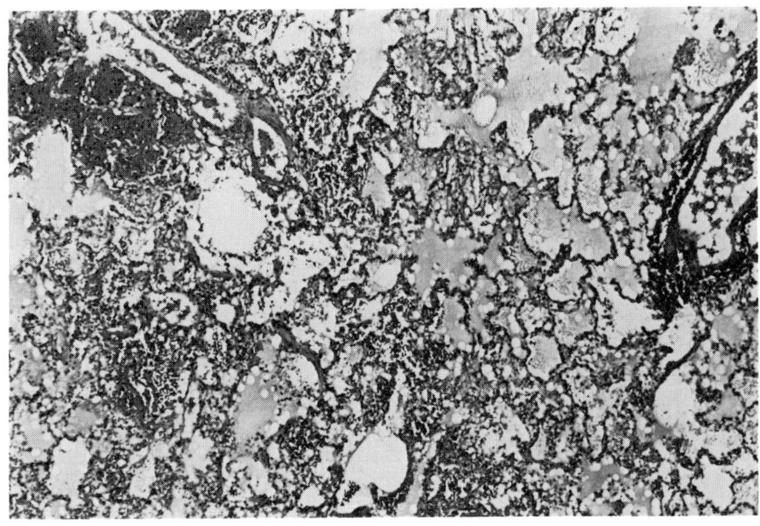

Fig. 9.   *Lung from a baboon dying at 49 hours. Note the diffuse alveolar congestion, hemorrhage, and edema. Hematoxylin and eosin, × 48. Reproduced from Archer et al (1983), with permission.*

Hemorrhages may be present but are not prominent (McGovern and Tiller 1980).

In studies by Archer et al (1983) and Coalson et al (1975, 1978a, 1979) of live *E. coli-* or endotoxin-induced shock in baboons (African-raised), no hemorrhage or ischemic enterocolitis of the gastrointestinal tract was observed. The only microscopic change found in the gastrointestinal tract of the baboons was ectasia of the villus tips. On the other hand, in work in progress by Hinshaw et al, gross post-mortem examinations of baboons (Oklahoma-raised) challenged with LD_{100} *E. coli* revealed that at least 30% of the nonsurvivors had hemorrhages in the gastrointestinal tract. Fifteen percent had hemorrhagic lesions the entire length of the gut, and the other 15% exhibited hemorrhagic lesions of the colon.

In sharp contrast, the dog in endotoxin-induced or live *E. coli*-induced shock exhibits massive hemorrhagic necrosis (Archer et al 1978b; Brunson et al 1966; Hinshaw 1968; Hinshaw et al 1970b, 1974b, 1977, 1979a; Reynolds et al 1972). Brunson et al (1966) oberved that 5–12 hours post-endotoxin hemorrhagic lesions of the small bowel began to appear in the dog and increased in severity during the next 12 hours. During the second 12-hour period, hemorrhages of the wall and lumen of the large bowel began to appear. The hemorrhagic lesions of the small and large bowel persisted or progressed after 24 hours. They stated that the intestinal lesions, although involving almost the entire small bowel in the early phases, were more severe in the ileum and later progressed to involve the large bowel. The difference in response to endotoxin or live organisms between the dog and nonhuman primate may be accounted for by: (a) hepatic venous constriction which causes massive pooling in the liver and splanchnic bed in the dog but not in the monkey (Brobmann et al 1970; Hinshaw 1968, 1972; Hinshaw et al 1970b, 1974c, 1979a; Kuida et al 1961; MacLean and Weil 1956; MacLean et al 1956; Weil et al 1956), and (b) vascular supply to the villus tips which are highly anastomotic and complex in the dog (Reynolds et al 1972). The villi shorten markedly in response to endotoxin, causing core and subepithelial vessels to be coiled and folded upon themselves (Reynolds et al 1972).

Necrotizing hemorrhagic enteritis is a hallmark of septic shock in the dog and is the major response that contributes to lethality in this model. In contrast, McGovern and Tiller (1980) state that the ischemic enterocolitis observed in patients contributes to the lethality of shock only when the condition is extensive. They state that reduction in perfusion below a critical level is the important factor in the production of these lesions. The level varies from site to site in the intestinal tract with the condition of the mesenteric vessels and the state of the vascular anastomoses. If there is severe hypotension, perfusion of the entire intestine may be reduced below the critical level, depending on the response of the intestine to catecholamines released as a result of the hypotension (McGovern and Tiller 1980).

Interestingly, McGovern and Tiller (1980) report that in the past few years ischemic enterocolitis as a manifestation of shock has been seen much less frequently than in previous years. This coincides with the less frequent use of vasopressors in the resuscitation of patients with septic shock and is particularly noticeable in cardiogenic shock (in which vasopressors are presumably uniformly used). This statement suggests that the ischemic enterocolitis observed in man during shock may not be so much a species response to shock but an iatrogenically exacerbated response to shock. However, in general, the response of the man during septic shock can lead to ischemic enterocolitic lesions that are seldom lethal (Marston et al 1966). The rapidity with which investigators generally induce shock by injecting (2–10 minutes) or infusing (up to 2 hours) endotoxin or bacteria may actually produce or exacerbate the gut lesion in the dog. It might be that investigators could design studies to avoid or lessen the hepatosplanchnic response of the dog by inducing sepsis gradually, for example with a prolonged infusion.

7.6. Heart

Morphologic lesions have been observed in hearts of patients who died from septic shock (Bohm 1982; Harms and Lehmann 1969; Masland and Barrows 1962; McGovern 1971; McGovern and Tiller 1980; Moon 1948; Remmele and Harms 1968; Robbins and Cotran 1979a; Sandritter and Lasch 1967). McGovern and Tiller (1980) observed increased numbers of PMN in the hearts of patients that died within two to three days after the onset of sepsis. When patients died five days after the onset of sepsis, lymphocytes, plasma cells, and Anitschkov cells had replaced the PMN. McGovern and Tiller (1980) stated that subendocardial hemorrhages and transverse tigroid stripes or bands of the heart were observed less often. Masland and Barrows (1962) reported that patients dying of shock associated with the general Shwartzman reaction had normal cardiac muscle fibers, but the small blood vessels of the heart showed proliferation and fibrin had adhered to the intimal surface. Many of these vessels were surrounded by PMN. Robbins and Cotran (1979a), summarizing morphologic lesions observed in the heart of patients after shock, stated that two fairly striking types of change appear: (a) subendocardial hemorrhages and necrosis, and (b) zonal lesions (Martin and Hackel 1963).

Coalson et al (1972) and Mela et al (1974) have documented accumulations of edema between and within myocardial myofibrils and in mitochondria during endotoxin shock in the dog. They reported that edema was severe enough in some cases to cause mitochondrial rupture.

Coalson et al also evaluated the hearts of baboons subjected to *E. coli* endotoxin (1978a) and live *E. coli* (1975, 1979) shock. They found myofibrils and mitochondria separated by edema within endocardial fibers, high amplitude swelling of mitochondria and increased numbers of contraction bands.

Discrete hemorrhages of epicardium have also been observed in nonsurviving baboons (Archer et al 1983). Myocardial hemorrhage, detected in the endotoxin-shocked dog 5 hours post-endotoxin, increased in severity from 5–12 hours post-endotoxin (Brunson et al 1966). In contrast, a minimum number of new cardiac lesions appeared during the next 12 hours.

7.7. Brain

Morphologic lesions have been observed in brain tissues from patients dying of septic shock (Lindqvist et al 1963; Masland and Barrows 1962; McKay et al 1959; McGovern 1971; McGovern and Tiller 1980; Sandritter and Lasch 1967). McGovern and Tiller (1980) state that the typical lesion in the brain is fibrin thrombi in venules surrounded by a ring of hemorrhage. The vessels may appear necrotic and a light PMN infiltrate is common. McKay et al (1959), Masland and Barrows (1962), and Lindqvist et al (1963), reporting cases of generalized Shwartzman phenomena associated with endotoxin shock, described the brain lesions observed as 'necrotizing encephalopathy'. Masland and Barrows (1962) described a lesion in the central white matter as follows: 'The small blood vessels in this region were uniformly affected. The capillaries characteristically showed endothelial proliferation and a majority had intraluminal thrombi. Numerous pericapillary hemorrhages were present. The venules were widely dilated and filled with thrombus which had a quasicrystalline appearance on hematoxylin and eosin stain. Many venules showed partial or complete necrosis of the vessel wall. The least affected venules were cuffed with round cells, while the more severely affected had polymorphonuclear cells mixed in and the most affected vessels had an inner region of hemorrhage surrounded by polymorphonuclear cells and round cells, while the adjacent parenchymal tissue was necrotic.'

They also stated that the tissue surrounding these vessels showed varying degrees of glial reaction. Many neurons had pyknotic nuclei, particularly the pyramidal cells. The normal laminar pattern appeared less distinct. As in other regions, the parenchymal changes were located circumferentially around diseased vessels with the most severe changes nearest the vessel. In general, microglial proliferation was minimal. However, many of the astroglial nuclei were swollen and numerous neuronal nuclei showed varying degrees of pyknosis (Masland and Barrows 1962).

Robbins and Cotran (1979c) state that during shock the brain may suffer so-called 'hypoxic encephalopathy'. The changes in the brain depend on the duration and intensity of the hypoxia and whether the patient survives long enough for the change to be manifested. In cases in which the patient survives only a few minutes or a few hours, no change is seen regardless of the severity of the insult. When the insult has been very slight, the nerve cells may recover function, and no anatomic change may occur. The first demonstrable anatomic change, seen after a survival of 12 to 24 hours, is in the nerve cells. Most vulnerable are the large cells in Sommer's sector of the hippocampus, followed by the Purkinje cells of the cerebellum.

Unfortunately, Archer et al (1983) did not examine the brain in their septic shock studies of baboons nor did Brunson et al (1966) in their study of endotoxin-shocked dogs.

8. PREVENTION OR REDUCTION OF MORPHOLOGIC LESIONS BY METHYL PREDNISOLONE SODIUM SUCCINATE PLUS GENTAMICIN SULFATE (*MPSS/GS*)

Archer et al (1983) have reported that when tissues of baboons subjected to lethal septic shock with or without MPSS/GS were analyzed by light microscopy, the survivors could be readily differentiated from the nonsurvivors, but the time of initiation of therapy could not be determined without decoding the slides. Delaying the time of initiation of treatment did not result in more severe tissue lesions or more organs involved. The extent of the lesions of the individual untreated baboons ranged from mild to massive, but all had multiple organ damage. When MPSS/GS treatment saved an animal from death, all significant organ damage was prevented or ameliorated. Sequential delay of initiation of therapy did not reveal an order in which organ damage was prevented by MPSS/GS. If all the animals had been killed at the same time (e.g., 12 hours), the probability of detecting any such preferential sequence of protection might have been greatly increased. However, we believe this would have more likely revealed the order of organ susceptibility to the pathophysiologic alterations initiated by *E. coli*. It is evident from evaluating the tissues of the protected baboons that MPSS/GS prevented or reduced the progressive development of lesions and allowed time for reparative processes of the animal. The data suggest that MPSS/GS works systemically to prevent or decrease the derangements in blood flow and blood volume of septic shock. MPSS/GS thus improves the cardiovascular status of an animal, and multiple organs are protected. Additionally, the success of MPSS/GS therapy may be related to the fact that MPSS probably aids in effectively distributing gentamicin to peripheral sites for the destruction of the organisms.

There are several current theories that might explain how the MPSS in MPSS/GS therapy might act to prevent or reduce the derangements in blood flow and blood volume. The MPSS may inhibit DIC (Balis et al 1978), complement-induced granulocyte margination, aggregation, and endothelial cell damage (Brigham et al 1981; Hammerschmidt et al 1979; Skubitz et al 1981), and thus prevent increased permeability and edema formation (Brigham et al 1981; Feola et al 1976; Sibbald et al 1981). The MPSS may block prostaglandin synthesis (Flower and Blackwell 1979; Gryglewski 1979). The MPSS may stabilize lysosomal membranes and prevent acid hydrolases from being released (Lefer and Martin 1969; Lefer and Verrier 1970). The MPSS may have vasodilatory and positive inotropic properties and thus increases venous return (Hinshaw et al 1967) and cardiac output (Dietzman et al 1973; Sambhi et al 1965), coronary blood flow (Hinshaw et al 1974a), cerebral blood flow (Emerson and Bryan 1977; Hinshaw et al 1967; Sullivan and Cavanagh 1966; Vaughn et al 1967), and augments myocardial contractility (Hinshaw et al 1974a; Tanz and Kerby 1961). Regardless of the mechanisms of action inherent in MPSS/GS combination therapy, the tissue results of this study unequivocably demonstrate that the morphologic manifestations of blood flow abnormalities (i.e., congestion, hemorrhage, edema, fibrin thrombi, etc.) were prevented or corrected in the surviving baboons.

8.1. Adrenal glands

Hemorrhage of the adrenal gland cortex was observed in 100% of the nonsurviving baboons (Fig. 1). The adrenal glands of the MPSS/GS-treated survivors (unlike nonsurvivors) were normal (Fig. 10) or had only mild to moderate residual congestion (Fig. 11). The consistency of this lesion suggests that the adrenal gland may be a key organ in the pathogenesis of irreversible shock. Both zona glomerulosa and zona fasciculata were most often affected and in some instances the zona reticularis was involved. The fasciculata mainly but also the reticularis produces the endogenous glucocorticoid cortisol. Administering an exogenous glucocorticoid (MPSS) increased survival rate in this 100% lethal model.

8.2. Liver

The liver lesions observed in 100% of the nonsurvivors: increased numbers of PMN (+1 to +3) and/or moderate to massive congestion (Fig. 2) were either not seen in survivors or were reduced in nature. In fact, the livers of 60% of the survivors appeared normal (Fig. 12) while livers of 40% showed residual mild congestion and/or +1 increase in numbers of PNM.

8.3. Kidneys

The lesions observed in the kidneys of nonsurvivors included moderate to severe cortical tubular necrosis (Fig. 4), moderate to massive congestion of cortical medullary and glomerular vessels, increased numbers of PMN (+1 to +3) within vessels and the interstitium, hemorrhage into the interstitium, glomerular fibrin thrombi (Fig. 6 and 7), and granular tubular casts and/or proteinuria (Fig. 5). In survivors these lesions were prevented or decreased. The kidneys of 60% of the survivors appeared normal (Fig. 13) while 40% had mild tubular necrosis and/or increased numbers of PMN (+1 to +2).

8.4. Lungs

The lung lesions in nonsurvivors ranging from increased numbers of PMN (+2 to +3) in the alveolar septa (Fig. 8), congestion (+2 to +3), interstitial edema (+1 to +3), focal or diffuse alveolar edema (+1 to +3), alveolar

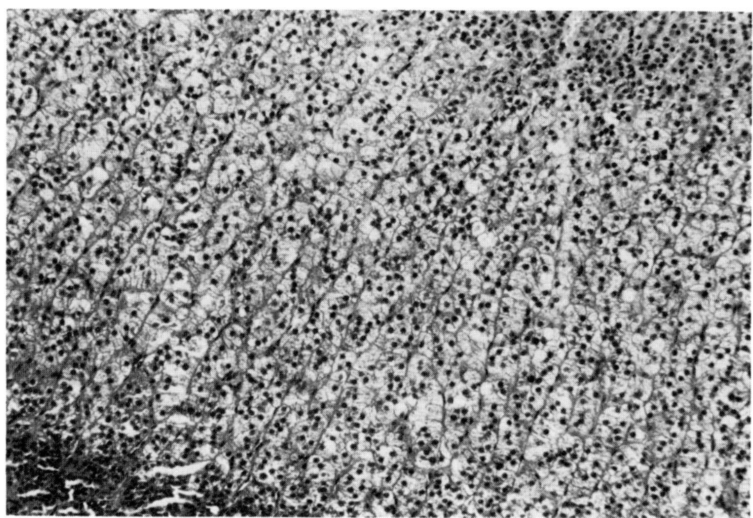

Fig. 10. *Adrenal gland representative of MPSS/GS-treated survivors in all three groups that were killed 14-71 days post-E. coli challenge. Cortical cells of both zonae glomerulosa and fasciculata appear normal. Hematoxylin and eosin, × 48. Reproduced from Archer et al (1983), with permission.*

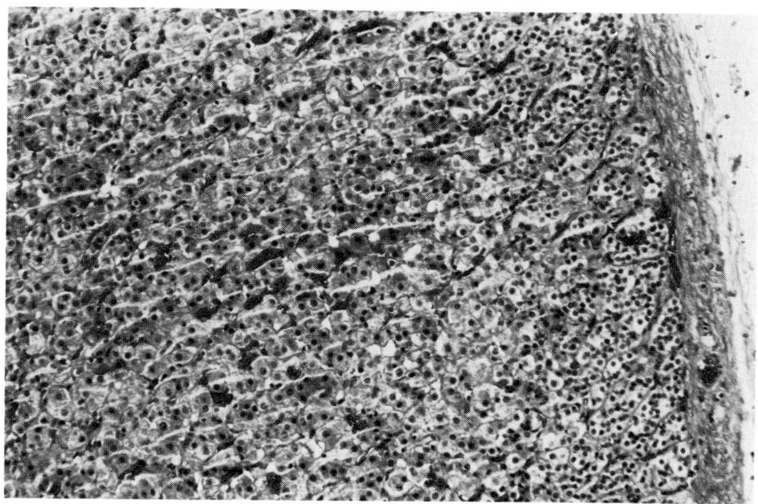

Fig. 11. *Adrenal gland representative of MPSS/GS-treated survivors in all three groups that were killed between 7 and 9 days post-E. coli challenge. Vessels in both zonae glomerulosa and fasciculata are congested while cortical cells appear normal. Hematoxylin and eosin, × 48. Reproduced from Archer et al (1983), with permission.*

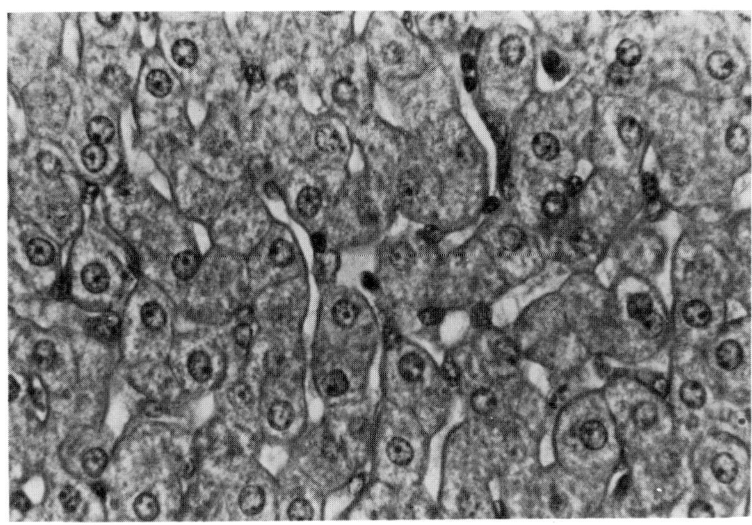

Fig. 12.   *Liver of MPSS/GS-treated survivor. Note lack of vascular congestion and PMN. Hematoxylin and eosin, × 192. Reproduced from Archer et al (1983), with permission.*

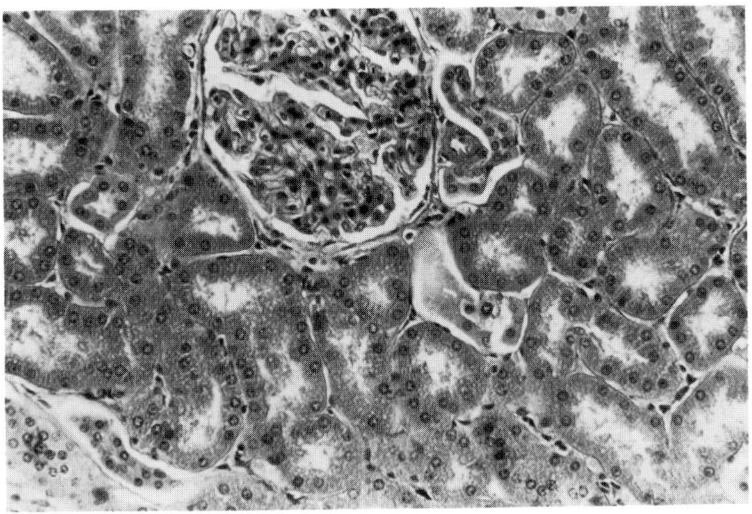

Fig. 13.   *Kidney of MPSS/GS-treated survivor. Glomeruli and tubules contain none of the changes described in the nonsurvivors. Hematoxylin and eosin, ×72. Reproduced from Archer et al (1983), with permission.*

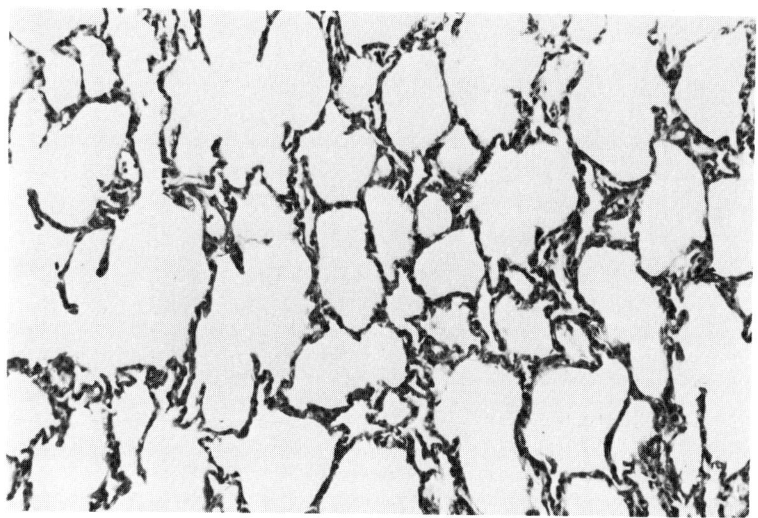

Fig. 14. *Lung from survivor. No significant lesions are present. Hematoxylin and eosin, × 48. Reproduced from Archer et al (1983), with permission.*

rounding, moderate hemorrhage, and fibrin thrombi to massive diffuse hemorrhage and edema of alveoli (Fig. 9) were greatly reduced in the survivors. Although congestion (+1 to +2) remained in the lungs of most survivors, the other lesions were not observed. Additionally, the lungs of three survivors appeared completely normal (Fig. 14).

9. CLINICAL RELEVANCE OF FINDINGS FROM STUDIES IN BABOONS

The results of histologic analysis demonstrate that MPSS/GS intermittent infusion therapy prevented or reduced morphologic lesions of LD_{100} *E. coli*-shocked baboons and increased the survival rate. The morphologic manifestations observed in untreated *E. coli*-shocked baboons are remarkably similar to those seen in tissues of patients who die in septic shock; this further substantiates the clinical relevance of the animal model. Finally, these findings lend strong support for the advisability of instituting early steroid/antibiotic therapy during severe sepsis in the human patient.

10.　SUMMARY

The knowledge derived from studying the morphologic derangements resulting from septic shock in various species provides a valuable tool for investigators to more definitively assess the effectiveness of different therapeutic regimens. Although the primary goal of a therapy study is to achieve permanent survival, the quality of life is also important. The evaluation of the effectiveness of a treatment regimen against the pathophysiologic effects of endotoxin/septic shock is therefore not complete until the tissues are analyzed and it is shown that lesions have been prevented or decreased. In conclusion, it is important to design septic shock therapy *studies* so that long-term survival benefits can be definitively established and the prevention or reduction of tissue damage can also be confirmed.

ACKNOWLEDGEMENTS

My sincere appreciation goes to Ms. Vickie Pennington and Ms. Kassie Hostetter for secretarial and editorial assistance, for their many hours spent, and for their effort beyond the call of duty.

REFERENCES

Altura BM Halevy S (1977) Circulatory shock, histamine and antihistamines: therapeutic aspects. In: Born GUR, Eichler O, Farah A, Herken H, Welch AD (Eds), *Handbuch der Experimentellen Pharmakologie*, p 575. Springer-Verlag, Berlin.

Archer LT, Black MR, Hinshaw LB (1975) Myocardial failure with altered response to adrenaline in endotoxin shock. *Br. J. Pharmacol. 54*, 145-155.

Archer LT, Beller BK, Drake JK, Whitsett TL, Hinshaw LB (1978a) Reversal of myocardial dysfunction in endotoxin shock with insulin. *Can J. Physiol. Pharm. 45*, 132-138.

Archer LT, White GL, Coalson JJ, Beller BK, Elmore O, Hinshaw LB (1978b) Preserved liver function and leukocyte response in superlethal endotoxic shock. *Circ. Shock 5*, 279-289.

Archer LT, Benjamin BA, Beller-Todd BK, Brackett DJ, Wilson MF, Hinshaw LB (1982) Does LD_{100} *E. coli* shock cause myocardial failure? *Circ. Shock 9*, 7-16.

Archer LT, Kosanke SD, Beller BK, Passey RB, Hinshaw LB (1983) Prevention or amelioration of morphologic lesions in LD_{100} *E. coli*-shocked baboons with steroid/antibiotic therapy. In: Reichard SM, Reynolds DG, Adams HR (Eds), *Advances in Shock Research Vol 10*, pp 195-215. Alan R. Liss, New York.

Balis JU, Gerber LI, Rappaport ES, Neville WE (1974) Mechanism of blood-vascular reactions of the primate lung to acute endotoxemia. *Exp. Mol. Pathol. 21*, 123-137.

Balis JU, Rappaport ES, Gerber L, Fareed J, Buddingh F, Messmore HL (1978) A primate model for prolonged endotoxin shock. Blood-vascular reactions and effects of glucocorticoid treatment. *Lab. Invest. 38*, 511-523.

Bergentz SE, Carlsten S, Gelin LE, Kreps J (1969) 'Hidden acidosis' in experimental shock. *Ann. Surg. 169*, 227-232.

Berne RM, Levy MN (1977) The microcirculation and lymphatics. In: *Cardiovascular Physiology, 3rd edition*, pp 119-126. C. V. Mosby, Saint Louis, MO.

Blaisdell FW (1974) Pathophysiology of the respiratory distress syndrome. *Arch. Surg. 108*, 44-49.

Bohm N (1982) Adrenal, cutaneous and myocardial lesions in fulminating endotoxemia (Waterhouse-Friderichsen syndrome). *Pathol. Res. Pract. 174*, 92-105.

Brigham KL, Bowers RE, McKeen CR (1981) Methylprednisolone prevention of increased lung vascular permeability following endotoxemia in sheep. *J. Clin. Invest. 67*, 1103-1110.

Brobmann GF, Ulano HB, Hinshaw LB, Jacobson ED (1970) Mesenteric vascular responses to endotoxin in the monkey and dog. *Am. J. Physiol. 219*, 1464-1467.

Brunson JG, Schulz DM, Angevine DM, Imrie ST, Douglas BH (1966) Evaluation of therapeutic agents: morphologic changes and survival data in dogs subjected to endotoxin during the shock tour. *J. Okla. State Med. Assoc. 59*, 479-484.

Cavanagh D, Rao PS, Comas MR (1977) Septic shock in obstetrics and gynecology. In: *Major Problems in Obstetrics and Gynecology Vol 11*, pp 1-129. W. B. Saunders Philadelphia.

Cerra FB, Siegel JH, Border JR, Peters DM, McMenamy RH (1979a) Correlations between metabolic and cardiopulmonary measurements in patients after trauma, general surgery and sepsis. *J. Trauma 19*, 621-629.

Cerra FB, Siegel JH, Border JR, Wiles J, McMenamy RH (1979b) The hepatic failure of sepsis: cellular versus substrate. *Surgery 86*, 409-422.

Cerra FB, Border JR, McMenamy RH, Siegel JH (1982) In: Cowley RA, Trump BV (Eds), *Pathophysiology of Shock, Anoxia and Ischemia*, pp 254-270. Williams and Wilkins, Baltimore.

Clowes GHA Jr, O'Donnell TF, Ryan NT, Blackburn GL (1974) Energy metabolism in sepsis: treatment based on different patterns in shock and high output stage. *Ann. Surg. 179*, 684-696.

Clowes GHA Jr, Martin H, Walji S, Hirsch E, Gazitua R, Goodfellow R (1978) Blood insulin responses to blood glucose levels in high output sepsis and septic shock. *Am. J. Surg. 135*, 577-583.

Coalson JJ, Woodruff HK, Greenfield LJ, Guenter CA, Hinshaw LB (1972) Effects of digoxin on myocardial ultrastructure in endotoxin shock. *Surg. Gynecol. Obstet.* *135*, 908-912.

Coalson JJ, Hinshaw LB, Guenter CA, Berrell EL, Greenfield LJ (1975) Pathophysiologic responses of the subhuman primate in experimental septic shock. *Lab. Invest. 32*, 561-569.

Coalson JJ, Benjamin B, Archer LT, Beller B, Gilliam CL, Taylor FB Jr, Hinshaw LB (1978a) Prolonged shock in the baboon subjected to infusion of *E. coli* endotoxin. *Circ. Shock 5*, 423-437.

Coalson JJ, Benjamin BA, Archer LT, Beller BK, Spaet RH, Hinshaw LB (1978b) A pathologic study of *Escherichia coli* shock in the baboon and the response to adrenocorticosteroid treatment. *Surg. Gynecol. Obstet. 147*, 726-736.

Coalson JJ, Archer LT, Benjamin BA, Beller-Todd BK, Hinshaw LB (1979) A morphologic study of live *Escherichia coli* organism shock in baboons. *Exp. Mol. Pathol. 31*, 10-22.

Cohn ZA, Hirsch JC (1960) The isolation and properties of the specific cytoplasmic granules of rabbit polymorphonuclear leukocytes. *J. Exp. Med. 112*, 983-1004.

Corrigan JJ, Ray WL, May N (1968) Changes in the blood coagulation system associated with septicemia. *N. Engl. J. Med. 279*, 851-856.

Cryer PE, Herman CM, Sode J (1971) Carbohydrate metabolism in the baboon subjected to gram-negative (*E. coli*) septicemia. I. Hyperglycemia with depressed plasma insulin concentrations. *Ann. Surg. 174*, 91-100.

Cryer PE, Herman CM, Sode J (1972) Lethal *Escherichia coli* septicemia in the baboon: alpha-adrenergic inhibition of insulin secretion and its relationship to the duration of survival. *J. Lab. Clin. Med. 79*, 622-638.

Dalgaard OZ (1960) An electron microscopic study on glomeruli in renal biopsies taken from human shock kidney. *Lab. Invest. 9*, 364-366.

Dietzman RH, Lillehei RC, Shatney CH (1973) Therapeutic effects of corticosteroids in septic shock. *Acta. Chir. Belg. 72*, 303-330.

Dinarello CA, Goldin NP, Wolff SM (1974) Demonstration and characterization of two distinct human leukocytic pyrogens. *J. Exp. Med. 139*, 1368-1381.

Dunnill MA (1974) A review of the pathology and pathogenesis of acute renal failure due to acute tubular necrosis. *J. Clin. Pathol. 27*, 2-13.

Emerson TE Jr, Bryan WJ (1977) Regional cerebral blood flows in endotoxin shock with methylprednisolone treatment. *Proc. Soc. Exp. Biol. Med. 156*, 378-381.

Feola M, Rovetto M, Soriano R, Cho SY, Wiener L (1976) Glucocorticoid protection of the myocardial cell membrane and the reduction of edema in experimental acute myocardial ischemia. *J. Thorac. Cardiovasc. Sug. 72*, 631-642.

Flower RJ, Blackwell GJ (1979) Anti-inflammatory steroids induce biosynthesis of a phospholipase A_2 inhibitor which prevents prostaglandin generation. *Nature (London) 278*, 456-459.

Friderichsen C (1918) Nebennierenapoplexie bei kleinen Kindern. *Jahrb. Kinderheilkd. 87*, 109-125.

Gilbert RP (1962) Endotoxin shock in the primate. *Proc. Soc. Exp. Biol. Med. 111*, 328-331.

Goldstein IM, Roos D, Kaplan HB, Weissman G (1975) Complement and immunoglobulins stimulate superoxide production by human leukocytes independently of phagocytosis. *J. Clin. Invest. 56*, 1155-1163.

Gordon S, Unkeless JC, Cohn ZA (1974) Induction of macrophage plasminogen activator by endotoxin stimulation and phagocytosis: evidence for a two-stage process. *J. Exp. Med. 140*, 995-1010.

Greendyke RM (1965) Adrenal hemorrhage. *Am. J. Clin. Pathol. 43*, 210-215.

Groves AC, Woolf LI, O'Regan PJ, Beach C, Hasinoff C, Sutherland WH (1974) Impaired gluconeogenesis in dogs with *E. coli* bacteremia. *Surgery 76*, 533-541.

Gryglewski RJ (1979) Effects of anti-inflammatory steroids on arachidonate cascade. In: Weissman G, Samuelsson B, Paoletti R (Eds), *Advances in Inflammation Research, Vol 1*, pp 505-513. Raven Press, New York.

Guntheroth WG, Jacky JP, Kawabori I, Stevenson JG, Moreno AH (1982) Left ventricular performance in endotoxin shock in dogs. *Am. J. Physiol. 242*, H172-H176.

Hamberg M, Svensson J, Samuelsson B (1975) Thromboxanes: a new group of biologically active compounds derived from prostaglandin endoperoxides. *Proc. Natl Acad. Sci. USA 72*, 2994-2998.

Hammerschmidt DE, White JC, Craddock PR, Jacob HS (1979) Corticosteroids inhibit complement-induced granulocyte aggregations. A possible mechanism for their efficacy in shock states. *J. Clin. Invest. 63*, 798-803.

Hardaway RM (1978) Acute respiratory distress syndrome and disseminated intravascular coagulation. *South Med. J. 71*, 596-598.

Harms D, Lehmann H (1969) Studies of peripheral microthrombosis in non-selected autopsy material. *Virchows Arch. A, 347*, 57-68.

Hassett J, Cerra FB, Siegel J, Moyer E, Yu L, Border JR, McMenamy R (1983) Symposium papers: multiple systems organ failure – a very brief summary. *Injury 14*, 93-97.

Hess ML, Kraus SM (1979) Myocardial subcellular function in shock. In: Crass MF III, Sordahl LA (Eds), Metabolic and morphologic correlates in cardiovascular function. *Texas Rep. Biol. Med. 39*, 193-207.

Hiller E, Saal JG, Ostendorf P, Griffiths GW (1977) The procoagulant activity of human granulocytes, lymphocytes and monocytes stimulated by endotoxin. *Klin. Wochenschr. 55*, 751-757.

Hinshaw LB (1968) Comparative effects of endotoxin on canine and primate intestine. *J. Surg. Res. 8*, 535-538.

Hinshaw LB (1971) Release of vasoactive agents and the vascular effects of endotoxin. In: Solomon K, Weinbaum G, Ajl SJ (Eds) *Microbial Toxins, Vol 5*, pp 209-260. Academic Press, New York.

Hinshaw LB (1972) Comparison of responses of canine and primate species to bacteria and bacterial endotoxin. In: Forscher BK, Lillehei RC, Stubbs SS (Eds), *Shock in Low- and High-Flow States*, pp 245-249. Excerpta Medica, Amsterdam.

Hinshaw LB (1982) Overview of endotoxin shock. In: Cowley RA, Trump BV (Eds), *Pathophysiology of Shock, Anoxia and Ischemia*, pp 219-235. Williams and Wilkins, Baltimore.

Hinshaw LB, Emerson TE Jr, Reins DA (1966) Cardiovascular responses of the primate in endotoxin shock. *Am. J. Physiol. 210*, 335-340.

Hinshaw LB, Solomon LA, Freeny PC, Reins DA (1967) Hemodynamic and survival effects of methylprednisolone in endotoxin shock. *Arch. Surg. 94*, 61-66.

Hinshaw LB, Mathis MC, Nanaeto JA, Holmes DD (1970a) Recovery patterns and lethal manifestations of live *E. coli* organism shock. *J. Trauma 10*, 787-794.

Hinshaw LB, Shanbour LL, Greenfield LJ, Coalson JJ (1970b) Mechanism of decreased venous return in subhuman primate administered endotoxin. *Arch. Surg. 100*, 600-606.

Hinshaw LB, Greenfield LJ, Owen SE, Black MR, Guenter CA (1972) Precipitation of cardiac failure in endotoxin shock. *Surg. Gynecol. Obstet. 135*, 39-48.

Hinshaw LB, Archer LT, Black MR, Greenfield LJ (1974a) Effects of methylprednisolone sodium succinate on myocardial performance, hemodynamic and metabolism in normal and failing hearts. In: Glenn TM (Ed), *Steroids and Shock*, pp 253-273. University Park Press, Baltimore, MD.

Hinshaw LB, Archer LT, Spitzer JJ, Black MR, Peyton MD, Greenfield LJ (1974b) Effects of coronary hypotension and endotoxin on myocardial performance. *Am. J. Physiol. 227*, 1051-1057.

Hinshaw LB, Peyton MD, Archer LT, Black MR, Coalson JJ, Greenfield LJ (1947c) Prevention of death in endotoxin shock by glucose administration. *Surg. Gynecol. Obstet. 139*, 851-859.

Hinshaw LB, Archer LT, Beller BK, White GL, Schroeder TM, Holmes DD (1977) Glucose utilization and the role of blood in endotoxin shock. *Am. J. Physiol. 233*, E71-E79.

Hinshaw LB, Coalson JJ, Benjamin BA, Archer LT, Beller BK, Kling OR, Hasser EM, Phillips RW (1978) *Escherichia coli* shock in the baboon and the response to adrenocorticosteroid treatment. *Surg. Gynecol. Obstet. 147*, 545-557.

Hinshaw LB, Beller BK, Archer LT, Flournoy DJ, White GL, Phillips RW (1979a) Recovery from lethal *Escherichia coli* shock in dogs. *Surg. Gynecol. Obstet. 149*, 545-553.

Hinshaw LB, Benjamin B, Archer LT, Peyton MD (1979b) The heart in endotoxin shock. *Texas Rep. Biol. Med. 39*, 173-191.

Hinshaw LB, Archer LT, Beller-Todd BK, Coalson JJ, Flournoy DJ, Passey R,

Benjamin B, White GL (1980) Survival of primates in LD$_{100}$ septic shock following steroid/antibiotic therapy. *J. Surg. Res. 28*, 151-170.

Hinshaw LB, Archer LT, Beller-Todd BK, Benjamin B, Flournoy DJ, Passey R (1981a) Survival of primates in lethal septic shock following delayed treatment with steroid. *Circ. Shock 8*, 291-300.

Hinshaw LB, Beller-Todd BK, Archer LT, Benjamin B, Flournoy DJ, Passey R, Wilson MF (1981b) Effectiveness of steroid/antibiotic treatment in primates administered LD$_{100}$ *Escherichia coli. Ann. Surg. 194*, 51-56.

Hinshaw LB, Brackett DJ, Archer LT, Beller BK, Wilson MF (1983) Detection of the hyperdynamic state of sepsis in the baboon during lethal *E. coli* infusion. *J. Trauma 23*, 361-365.

Hinshaw LB, Beller BK, Archer LT, Schumer W (1985) Septic shock. In: Hardaway RM (Ed), *Shock: the Reversible Step Toward Death*. John Wright, PSG Inc. In press.

Hoyer JR, Seiler MW (1979) Pathophysiology of Tamm-Horsfall protein. *Kidney Int. 16*, 279-289.

Janoff A (1964) Alterations in lysosomes (intracellular enzymes) during shock; effects of preconditioning (tolerance) and protective drugs. In: Hershey SG (Ed), *Shock*, pp 93-111. Little, Brown and Co., Boston.

Jubiz W, Radmark O, Malmsten C, Hansson G, Lindgren JA, Plamblad J, Uden AM, Samuelsson B (1982) A novel leukotriene produced by stimulation of leukocytes with formylmethionyl-leucylphenylalanine. *J. Biol. Chem. 257*, 6106-6110.

Kobold E, Lovell R, Katz W, Thal AP (1964) Chemical mediators released by endotoxin. *Surg. Gynecol. Obstet. 118*, 807-813.

Kosanke SD, White GL, Archer LT, Britz WE Jr (1982) A comparison of the morphologic lesions in the dog and baboon septic shock models. *Lab. Anim. Sci. 32*, 420.

Kreger BE, Craven DE, McCabe WR (1980) Gram-negative bacteremia. IV. Reevaluation of clinical features and treatment in 612 patients. *Am. J. Med. 68*, 344-355.

Kuida J, Gilbert RP, Hinshaw LB, Brunson JG, Visscher MB (1961) Species differences in effect of gram-negative endotoxin on circulation. *Am. J. Physiol. 200*, 1197-1202.

Kurland JL, Bockman R (1978) Prostaglandin E production by human blood monocytes and mouse peritoneal macrophages. *J. Exp. Med. 147*, 952-957.

Lefer AM, Martin J (1969) Mechanism of the protective effect of corticosteroids in hemorrhagic shock. *Am. J. Physiol. 216*, 314-320.

Lefer AM, Verrier RL (1970) Role of corticosteroids in the treatment of circulatory collapse states. *Clin. Pharmacol. Ther. 11*, 630-655.

Lillehei RC, MacLean LD (1959) Physiological approach to successful treatment of endotoxin shock in the experimental animal. *Arch. Surg. 78*, 464-471.

Lindqvist B, Erlanson P, Brun A (1963) A case of renal cortical necrosis probably caused by a human equivalent of the Shwartzman reaction. *Acta Med. Scand. 173*, 561-567.

MacLean LD, Weil MH (1956) Hypotension (shock) in dogs produced by *Escherichia coli* endotoxin. *Circ. Res. 4*, 546-556.

MacLean LD, Weil MH, Spink WW, Visscher MB (1956) Canine intestinal and liver weight changes induced by *E. coli* endotoxin. *Proc. Soc. Exp. Biol. Med. 92*, 602-605.

MacLean LD, Duff JH, Scott HM, Peretz DI (1965) Treatment of shock in man based on hemodynamic diagnosis. *Surg. Gynecol. Obstet. 120*, 1-16.

MacLean LD, Mulligan WG, McLean APH, Duff JH (1967) Patterns of septic shock in man – a detailed study of 56 patients. *Ann. Surg. 166*, 543-562.

Marston A (1962) The bowel in shock. *Lancet 2*, 365-370.

Marston A, Pheils MT, Thomas ML, Morson BC (1966) Ischaemic colitis. *Gut 7*, 1-15.

Martin AM Jr, Hackel DB (1963) The myocardium of the dog in hemorrhagic shock. A histochemical study. *Lab. Invest. 12*, 77-91.

Martland HS, (1944) Fulminating meningococci infection with bilateral massive adrenal hemorrhage (the Waterhouse-Friderichsen syndrome). *Arch. Pathol. 37*, 147-158.

Masland WS, Barrows LJ (1962) A case of generalized Shwartzman phenomenon in the human. *Arch. Neurol. (Chicago) 7*, 64-73.

Mason JW, Colman RW (1971) The role of Hageman factor in disseminated intravascular coagulation induced by septicemia, neoplasia or liver disease. *Thromb. Diath. Haemorrh. 26*, 325-331.

McGovern VJ (1971) Shock. In: Sommers S (Ed), *Pathology Annual. Vol 6*, pp 279-298. Appleton-Century-Crofts, New York.

McGovern VJ, Tiller DJ (1980) *Shock: A Clinicopathologic Correlation.* Masson Publishing USA, New York.

McKay DG, Jewett JF, Reid DE (1959) Endotoxin shock and the generalized Shwartzman reaction in pregnancy. *Am. J. Obstet. Gynecol. 78*, 546-566.

McKay DG, Margaretten W, Csavossy I (1967) An electron microscope study of endotoxin shock in rhesus monkeys. *Surg. Gynecol. Obstet. 125*, 825-832.

Mela L, Bacalzo LV Jr, Miller LD (1971) Defective oxidative metabolism of rat liver mitochondria in hemorrhagic and endotoxin shock. *Am. J. Physiol. 220*, 571-577.

Mela L, Hinshaw LB, Coalson JJ (1974) Correlation of cardiac performance, ultrastructural morphology and mitochondrial function in endotoxemia in the dog. *Circ. Shock I*, 265-272.

Milligan GF, MacDonald JAE, Melton A, Ledingham IMA (1974) Pulmonary and hematologic disturbances during septic shock. *Surg. Gynecol. Obstet. 138*, 43-49.

Moon VH (1948) The pathology of secondary shock. *Am. J. Pathol. 24*, 235-273.

Morrison DC, Ulevitch RJ (1978) The effects of bacterial endotoxins on host mediation systems. *Am. J. Pathol. 93*, 527-617.

Nies AS, Forsyth RP, Williams HE, Melmon KL (1968) Contribution of kinins to endotoxin shock in unanesthetized rhesus monkeys. *Circ. Res. 22*, 155-164.

Nishijima H, Weil MH, Shubin H, Cavanilles J (1973) Hemodynamic and metabolic studies on shock associated with gram-negative bacteremia. *Medicine 52*, 287-294.

Orell SR (1971) Lung pathology in respiratory distress following shock in the adult. *Acta. Pathol. Microbiol. Scand. Sect. A Pathol. 79*, 65-76.

Powrie S, Norman J (1976) Septicaemia. *Br. J. Anaesth. 48*, 41-47.

Remmele W, Harms D (1968) Pathologic anatomy of circulatory shock in man: I. Microthrombosis of peripheral blood vessels. *Klin. Wochenschr. 46*, 352-357.

Reynolds DG, Brungardt JM, Swan KB (1972) Effects of endotoxin on the vascular architecture of intestinal mucosa. In: Hinshaw LB, Cox BG (Eds), *The Fundamental Mechanisms of Shock*, pp 113-126. Plenum Press, New York.

Rich AR (1944) Peculiar type of adrenal cortical damage associated with acute infection and its possible relation to circulatory collapse. *Bull. Johns Hopkins Hosp. 74*, 1-15.

Riede UN, Joachim H, Costabel U, Hassenstein J, Sandritter W, Augustin P, Mittermayer Ch (1978) The pulmonary air-blood barrier of human shock lung (a clinical ultrastructural and morphometric study). *Pathol. Res. Pract. 162*, 41-72.

Riede UN, Mittermeyer Ch, Horn R, Sandritter W (1981) Pathobiology of the alveolar wall in human shock lung. In: Cowley RA, Trump BF (Eds), *Pathophysiology of Shock, Anoxia and Ischemia*, pp 358-371. Williams and Wilkins, Baltimore.

Richman AV, Gerber LI, Balis JU (1980) Peritubular capillaries. A major target site of endotoxin-induced vascular injury in the primate kidney. *Lab. Invest. 43*, 327-332.

Rietschel ET, Schade U, Jensen M, Wollenweber H-W, Lüderitz O, Greisman SG (1982) Bacterial endotoxins: chemical structure, biologic activity and role in septicaemia. *Scand. J. Infect. Dis. (Suppl) 31*, 8-21.

Rivers RPA, Hathaway WE, Weston WL (1975) The endotoxin-induced coagulant activity of human monocytes. *Br. J. Haematol. 30*, 311-316.

Robbins SL, Cotran RS (1979a) Fluid and hemodynamic derangements. In: *Pathologic Basis of Disease, 2nd edition*, pp 107-140. W. B. Saunders, Philadelphia, PA.

Robbins SL, Cotran RS (1979b) The kidney. In: *Pathologic Basis of Disease, 2nd edition*, pp 1149-1150. W.B. Saunders, Philadelphia, PA.

Robbins SL, Cotran RS (1979c) The nervous system. In: *Pathologic Basis of Disease, 2nd edition*, pp 1552-1553. W.B. Saunders, Philadelphia, PA.

Robboy SJ, Colman RW, Minna JD (1972) Pathology of disseminated intravascular coagulation (DIC): analysis of 26 cases. *Hum. Pathol. 3*, 327-343.

Robinson JA, Klodnycky ML, Loeb HS, Racic MR, Gunnar RM (1975) Endotoxin, prekallikrein, complement and systemic vascular resistance: sequential measurements in man. *Am. J. Med. 59*, 61-67.

Sambhi MP, Weil MH, Udhoji VN (1965) Acute pharmacodynamic effects of glucocorticoids: cardiac output and related hemodynamic changes in normal subjects and patients in shock. *Circulation 31*, 523-530.

Sandritter W, Lasch HG (1967) Pathologic aspects of shock. In: Bajusz E, Jasmin G (Eds), *Methods and Achievement in Experimental Pathology, Vol 3*, pp 86-121. Karger, Basel-New York.

Sibbald WJ (1976) Bacteremia and endotoxemia: a discussion of their roles in the pathophysiology of gram-negative sepsis. *Heart Lung 5*, 765-771.

Sibbald WJ, Anderson RR, Reid B, Holliday RL, Dreidger AA (1981) Alveolocapillary permeability in human septic adult respiratory distress syndrome: effect of high-dose corticosteroid therapy. *Chest 79*, 133-142.

Siegel JH, Greenspan M, Del Guercio LRM (1967) Abnormal vascular tone, defective oxygen transport, and myocardial failure in human septic shock. *Ann. Surg. 165*, 504-517.

Siegel JH, Goldwyn RM, Friedman HP (1971) Pattern and process in the evolution of human septic shock. *Surgery 70*, 232-245.

Siegel JH, Giovannini I, Coleman B (1979) Ventilation: perfusion maldistribution secondary to the hyperdynamic cardiovascular state as the major cause of increased pulmonary shunting in human sepsis. *J. Trauma 19*, 432-460.

Skubitz KM, Craddock PR, Hammerschmidt DE, August JT (1981) Corticosteroids block binding of chemotactic peptide to its receptor on granulocytes and cause disaggregation of granulocyte aggregates *in vitro. J. Clin. Invest. 68*, 13-20.

Slater TF (1969) Lysosomes and experimentally induced tissue injury. In: Dingle JT, Fell HB (Eds), *Lysosomes in Biology and Pathology*, p 469. Elsevier, Amsterdam-New York.

Spink WW (1972) The role of endotoxin in shock. In: Forscher BK, Lillehei RC, Stubbs SS (Eds), *Shock in Low- and High-Flow States*, pp 226-229. Excerpta Medica, Amsterdam.

Spink WW, Braude AI, Castaneda MR, Groytea RS (1948) Auromycin therapy in human brucellosis due to *Brucella melitensis. J. Am. Med. Assoc. 138*, 1145-1148.

Sullivan TJ III, Cavanagh D (1966) Corticosteroids in endotoxin shock: effect on renal vasomotion. *Arch. Surg. 92*, 732-739.

Tanz RD, Kerby CF (1961) The inotropic action of certain steroids upon isolated cardiac tissue; with comments on steroidal cardiotonic structure-activity relationships. *J. Pharmacol. Exp. Ther. 131*, 56-64.

Thal AP (1972) The cardiovascular response to sepsis. In: Forscher BK, Lillehei RC, Stubbs SS (Eds), *Shock in Low- and High-Flow States*, pp 240-244. Excerpta Medica, Amsterdam.

Trump BF (1975) The application of electron microscopy and cellular biochemistry to the autopsy, observations on cellular changes in human shock. *Human Pathol. 6*, 499-516.

Udhoji VM, Weil M (1965) Hemodynamic and metabolic studies on shock associated with bacteremia. *Ann. Intern. Med. 62*, 966-978.

Udhoji VA, Weil MH, Sambhi MP, Rosoff L (1963) Hemodynamic studies on clinical shock associated with infection. *Am. J. Med. 34*, 461-469.

Unanue ER (1976) Secretory function of mononuclear phagocytes: a review. *Am. J. Pathol. 83*, 396-417.

Vassalli P, Richet G (1961) Nécrose corticale et insuffisance rénal aiguë des états de choc. In: Richet EPG (Ed), *Proceedings of the First International Congress of Nephrology*, pp 236-320. Karger, Basel-New York.

Vaughn DT, Kirschbaum TH, Bersentes T, Assali NS (1967) Effects of corticosteroid hormones on regional circulation in endotoxin shock. *Proc. Soc. Exp. Biol. Med. 123*, 760-764.

Visscher MB (1972) An overview of the shock problem. In: Hinshaw LB, Cox BG (Eds), *The Fundamental Mechanisms of Shock*, pp 3-7. Plenum Press, New York.

Wahl LM, Wahl SM, Mergenhagen SR, Martin GR (1974) Collagenase production by endotoxin-activated macrophages. *Proc. Natl Acad. Sci. USA 71*, 3598-3601.

Waisbren BA (1964) Gram-negative shock and endotoxin shock. *Am. J. Med. 36*, 819-824.

Waterhouse R (1911) A case of suprarenal apoplexy. *Lancet 1*, 577-578.

Weil MH (1972) Proposed classifications of shock states, with special reference to distributive defects. In: Hinshaw LB, Cox BG (Eds), *The Fundamental Mechanisms of Shock*, pp 13-22. Plenum Press, New York.

Weil MH (1977) Current understanding of mechanisms and treatment of circulatory shock caused by bacterial infections. *Ann. Clin. Res. 9*, 181-191.

Weil MH, MacLean LD, Visscher MB, Spink WW (1956) Studies on the circulatory changes in the dog produced by endotoxin from gram-negative micro-organisms. *J. Clin. Invest. 35*, 1191-1198.

Weisel RD, Vito L, Dennis RC, Valeri CR, Hechtman HB (1977) Myocardial depression during sepsis. *Am. J. Surg. 133*, 512-522.

Weisul JP, O'Donnell TE Jr, Stone MA, Clowes GHA (1975) Myocardial performance in clinical septic shock: effects of isoproterenol and glucose potassium insulin. *J. Surg. Res. 18*, 357-363.

Wilson RF (1976) The diagnosis and management of severe sepsis and septic shock. *Heart Lung 5*, 422-429.

Wilson RF (1980) The pathophysiology of shock. *Intensive Care Med. 6*, 89-100.

Xarli VP, Steele AA, Davis PJ, Buescher ES, Rios CN, Garcia-Bunuel R (1978) Adrenal hemorrhage in the adult. *Medicine 57*, 211-221.

Handbook of Endotoxin, Vol. 4: Clinical Aspects of Endotoxin Shock
R.A. Proctor, editor
© Elsevier Science Publishers B.V., 1986

CHAPTER 3

Endotoxin shock in man caused by gram-negative bacilli*

Etiology, clinical features, diagnosis, natural history, and prevention

RICHARD J. HAMILL AND DENNIS G. MAKI

1. BACKGROUND AND DEFINITIONS

No bacterial infection with the possible exception of meningitis evokes greater fear and respect than bacteremia, particularly when complicated by the syndrome of endotoxic shock. Life-threatening bacterial infections of all types typically express themselves early with invasion of the bloodstream. Despite major advances in almost every aspect of therapeutic medicine and the availability of a burgeoning array of potent antimicrobial drugs effective against virtually all known human bacterial pathogens, half of the patients developing gram-negative bacteremia** with endotoxic shock die as a direct or indirect consequence of the infection. A case of gram-negative bacteremia constitutes a clinical emergency; a cluster of nosocomial cases, an even greater emergency, both clinically and epidemiologically. Over the past 30 years, the incidence of these infections has increased significantly in the developed countries. Fortunately much has been learned about the biology and clinical

---

* Portions of this review are excerpted, with permission, from a monograph 'Epidemic Nosocomial Bacteremia' written by one of the authors, D.G. Maki, and published in *Handbook of Hospital Infections* (R.P. Wenzel, editor), CRC Press, Boca Raton, Florida, 1981.

** In this review, the term 'gram-negative bacteremia' refers to bacteremia caused by gram-negative *bacilli*, that is, gram-negative bacillemia.

aspects of gram-negative sepsis and endotoxic shock, and about the epidemiology of bacteremic gram-negative infections in the hospital.

The goals of this chapter are: to characterize the clinical features of gram-negative sepsis, particularly as associated with endotoxin shock; to review current concepts of the laboratory diagnosis of gram-negative sepsis; to define the major sources of gram-negative bacteremia associated with endotoxic shock; to delineate risk factors predisposing to gram-negative bacteremia, endotoxic shock, and a fatal outcome; to examine the natural history of gram-negative sepsis with endotoxic shock and secular trends of gram-negative bacteremia; and to review briefly current measures for the prevention of gram-negative bacillary infections that produce bacteremia and endotoxic shock.

1.1. Definitions

The medical literature on gram-negative septicemia and endotoxic shock uses a number of terms, too often without clearly defining their precise meaning. Maki (1981) has offered the following definitions in an attempt to bring greater precision to clinical discussions of this complex subject:

Bacteremia is defined as the presence of bacteria in the bloodstream, as documented by culture or by histopathologic examination. Whereas it is clear that many bacteremias, especially when transient such as those arising from dental extractions, are clinically asymptomatic, gram-negative bacteremia is usually associated with clinical signs and symptoms (e.g., fever, chills, or hypotension) or laboratory evidence (e.g., leukocytosis or acid–base abnormalities) suggestive of bloodstream infection; clinically symptomatic bacteremia is referred to as *septicemia*. In this chapter, the two terms will be used interchangeably.

Sepsis is a less precise term which refers to a constellation of clinical signs and symptoms, such as fever, rigors, hypotension or even shock, that suggest systemic infection, in particular bacteremia. However, some patients with an obvious local infection (e.g., peritonitis or gram-negative pneumonia) and a clinical picture indiscernable from culture-proved gram-negative bacteremia will have negative blood cultures; some may even have full-blown endotoxic shock. The term *gram-negative sepsis*, with or without shock, applies to these patients as much as to patients with culture-documented gram-negative bacteremia (Koliner et al 1980).

Not all blood cultures showing microbial growth represent true bacteremia, the presence of those same microorganisms in the patient's bloodstream. Approximately 5% of positive blood cultures in most hospitals are ultimately judged to be contaminated by microorganisms introduced during the drawing of the blood culture or during its processing in the laboratory. In most of these cases, only one of multiple blood cultures is positive (Weinstein et al 1983a), usually with a skin commensal such as *Staphylococcus epidermidis*, *Bacillus* species, or a diphtheroid, in which instance it is easily recognized as a contaminant. If, on the other hand, multiple blood cultures from the same patient are positive, the relevance of the positive cultures must be established on clinical and epidemiologic grounds; such judgements can occasionally be difficult. Clusters of contaminated blood cultures have given rise to the term *epidemic pseudobacteremia* (Maki 1980).

Local infections involving vascular endothelium or the endocardium, usually associated with bacteremia, are termed *endovascular infections*. Bacteremia can originate from primary endovascular foci of infection (e.g., endocarditis or a vascular cannula) or, if originating from a nonvascular focus of infection, can subsequently produce endovascular infection (e.g., endocarditis). Endocarditis is a local infection of the lining of the heart, usually, but not invariably, involving one or more heart valves. Similarly, infection of a vein segment, documented histopathologically, is called *septic* (or *suppurative*) *phlebitis* and of an artery, *septic endarteritis*.

Nosocomial infections are infections that develop in the hospital and were not incubating at the time of admission.

The National Nosocomial Infection Surveillance study of the U.S. Centers for Disease Control (CDC) subdivides nosocomial bacteremias into two categories, *primary bacteremias* and *secondary bacteremias*. Primary bacteremias are culture-documented bacteremias occurring in the absence of a recognized infection with the responsible pathogen at a local anatomic site, such as the urinary tract or a surgical wound. Cases of bacteremia thought to be related to infusion therapy, in the absence of purulent (suppurative) thrombophlebitis, are also defined as primary. Conversely, bacteremia originated from culture-documented infection at a local anatomic site (e.g., the urinary tract) is regarded as a secondary bacteremia, and bacteremias deriving from culture-proved purulent thrombophlebitis are also considered secondary bacteremias. This categorization is admittedly arbitrary, but necessary under the constrictions of data submitted on many cases reported to the CDC. Whereas bacteremia occurring in severely compromised patients such as individuals with profound granulocytopenia is often cryptogenic, it is usually possible in most nosocomial bacteremias to identify the local infection which gave rise to the bacteremia. Failure to obtain complete cultures obviously does not imply that a local source for a bacteremia does not exist.

The most grave sequel of major infections of all types is the syndrome of *septic shock*. A wide variety of microorganisms besides gram-negative bacteria, including gram-positive bacteria, spirochetes, mycoplasmas, rickettsiae, chlamydia, fungi, viruses, and even parasites, have been implicated as causes of septic shock; however, this review will concentrate on bacteremic sepsis and the syndrome of septic or *endotoxic shock* caused by gram-negative bacilli. Gram-negative bacilli are by far the most frequent cause of septic shock, not only because a major portion of their bacterial cell wall consists of endotoxin, but because of their biologic ubiquity and predilection to cause systemic infection in a wide spectrum of clinical settings.

Hundreds of papers and monographs have been written on the subject of septic (or endotoxic) shock, and there are nearly as many definitions of this syndrome. We suggest that the following criteria be fulfilled before making a diagnosis of endotoxic shock: (a) documented gram-negative septicemia or, if bacteremia is not present, a symptomatic *major* local gram-negative infection – that is, a clear-cut picture of gram-negative sepsis, (b) evidence of generalized hypoperfusion, almost always – but not invariably – associated with hypotension (systolic blood pressure below 90 mmHg, or 50 mm less than a previously identified stable blood pressure in a known hypertensive patient). Hypoperfusion is better characterized by lactemia and metabolic acidosis and signs of organ failure from inadequate perfusion, including altered mental status, oliguria, a high (or low) cardiac output associated with an incommensurately low systemic vascular resistance, noncardiogenic pulmonary edema, ileus, or coagulopathies, particularly disseminated intravascular coagulation.

Whereas many authors have regarded hypotension occurring in the setting of gram-negative bacteremia as synonymous with endotoxic shock, it is very important to recognize that many patients with gram-negative sepsis and hypotension are not in shock but merely have relative hypovolemia, which is easily treated with adequate intravenous fluid replacement. The syndrome of endotoxic shock, which is discussed at length by other authors in these volumes (see Chapters 1, 2 and 6 and Volume 2), is a highly complex biologic entity thought to derive from gram-negative endotoxin and associated with a diverse array of host responses, resulting in generalized cellular hypoperfusion and auguring a grave prognosis for survival of the patient. Until reliable biochemical markers for this syndrome become clinically available, we believe that clinicians as well as investigators would do better to put greater emphasis on identifying clear-cut organ hypoperfusion in association with gram-negative sepsis, rather than hypotension alone, in defining the syndrome of endotoxic shock.

1.2. History

Until the 1950s, gram-negative bacteremia and associated shock were only rarely addressed in the medical literature. Through 1909, Jacob was able to find only 39 reported cases of documented *Escherichia coli* bacteremia in a review of the world literature (Jacob 1909). In 1924, Felty and Keefer reported 28 cases of *E. coli* bloodstream infection that had occurred in patients in the Johns Hopkins Hospital (Felty and Keefer 1924).

The observations and conclusions of these early authors hold true today; the most common portals of entry for *E. coli* bacteremia are the urinary tract, female genital infections, and other intraabdominal foci of infection. Urologic manipulations and surgery are frequent predisposing factors for bloodstream invasion. Case-fatality rates of *E. coli* bacteremia in this early pre-antibiotic period were surprisingly low, only about 32%, which is comparable to rates recorded today.

Subsequent descriptions of gram-negative bacteremia consisted primarily of isolated case reports until 1951 when Waisbren summarized the features and natural history of 29 episodes of gram-negative bacteremia occurring in patients in the University of Minnesota Hospitals (Waisbren 1951). Waisbren discerned two distinct clinical syndromes: he interpreted one group of patients' clinical course as reflecting a 'toxic' reaction to an invading gram-negative bacillus, whereas a second group's illness, often associated with shock, was thought to represent the systemic effects of bacterial endotoxin. Waisbren speculated that any gram-negative bacillus was capable of producing either or both syndromes. Moreover, he was the first to appreciate the enormous influence of the patient's underlying disease on the natural course and outcome of gram-negative bacteremia and accurately predicted the now well-documented evolution of bacterial species then thought to be non-pathogenic (e.g., *Pseudomonas aeruginosa*) into major human pathogens.

In 1959, Finland and his coworkers published the first of a series of monographs detailing secular trends in the incidence and natural history of bacteremic and other life-threatening bacterial infections encountered in the Boston City Hospital over a 40-year period, beginning in the pre-antibiotic era and continuing through the mid 1970s, a period of very heavy antibiotic use in U.S. hospitals (Finland et al 1959; McGowan et al 1975). These workers were among the first to point up a direct association between the rising frequency of gram-negative bacteremia and increasingly heavy use of systemic antibiotics in hospitals. Subjects not addressed by Finland and his coworkers of which we now know that they have probably contributed even more to the increasing problem with gram-negative infections of all types, particularly those originating in the hospital, were the increasing use of invasive devices and therapeutic procedures and of antineoplastic drugs and immunosuppressive therapy during this same period.

In 1962, McCabe and Jackson published a now classic monograph on the subject of gram-negative bacteremia, reporting secular trends in the incidence of gram-negative bloodstream infections seen in the University of Illinois Hospital between 1951 and 1959, the microbiologic, clinical and laboratory features of infection associated with 173 documented cases, and the outcome of therapy (McCabe and Jackson 1962a, 1962b). These authors were the first to quantify the profound effect of severity of underlying diseases on the outcome of gram-negative bacteremic infection. McCabe and Jackson described a simple but novel system for classifying the severity of patients' underlying diseases. By this system, patients not expected to survive more than one year under any circumstances (mainly those with acute leukemia) were classified as having a 'rapidly fatal disease'; patients whose underlying disease was of such severity that it would likely prove fatal during the next 4 years (e.g., metastatic carcinoma, lymphoma, cirrhosis with bleeding varices) were classified as having an 'ultimately fatal disease'; patients with diseases not expected to prove fatal during the next 4 years were classified as having 'nonfatal diseases'. Applying this categorization to their cases, McCabe and Jackson showed, and numerous other investigators have subsequently confirmed (Bryan et al 1983b; Scheckler 1977; Weinstein et al 1983b), that the nature and severity of a bacteremic patient's underlying condition is a highly reliable prognostic index: in their patients with rapidly fatal diseases, the mortality was 84%; in patients with ultimately fatal diseases, 48%, and in patients with nonfatal diseases, 16%.

Subsequent reports have further clarified the epidemiology, ecology, clinical features, and natural history of gram-negative bacteremia (Bryan et al 1983b; Dupont and Spink 1969; Kreger et al 1980a, 1980b; Lewis and Fekety 1969; Maiztegui et al 1965; Maki 1981; McGowan et al 1975; McHenry et al 1975; Myerowitz et al 1971; Scheckler 1977; Setia and Gross 1977; Weil et al 1964; Weinstein et al 1983a, 1983b). In particular, the enormous impact of various invasive devices and procedures which are now used routinely in modern hospitals for diagnosis and therapy on susceptibility to gram-negative bacillary infections of all types, and the considerable impact of nosocomial acquisition on morbidity and mortality have been well defined. A vast literature has arisen in recent years, reviewed elsewhere in these volumes, in which the biology of gram-negative bacillary infection, the biochemical events occurring with endotoxic shock and the efficacy of novel therapeutic modalities, beyond anti-infective drugs, are discussed.

Among the most exciting recent advances in our understanding of the biology of endotoxic shock and therapeutic application of this research has been a line of investigation directed at the use of immunotherapy, both for prevention and for treatment of gram-negative bacillary infections. As discussed in Chapter 5, in the early 1970s it was shown that the core lipopolysaccharide of the gram-negative cell wall, consisting primarily of N-acetylglucosamine, 3-deoxy-D-*manno*-2-octulosonic acid (KDO), heptose and glucose, linked to lipid A, appears to be shared by most of the Enterobacteriaceae, *Haemophilus influenzae* type b, and *P. aeruginosa* as well, is highly immunogenic and in animal models stimulates the production of antibodies which confer protection against lethal gram-negative infection (McCabe 1972; Young et al 1975; Ziegler et al 1973). In 1972, McCabe et al reported that pre-existent antibody to the core antigen correlated with a significantly improved outcome in patients with gram-negative bacteremia (McCabe et al 1972). More recently, Ziegler and her coworkers (1982) have shown in a prospective, double-blind clinical trial that administration of an antiserum with a high titer of antibody to the core antigen, derived by immunization of human volunteers with a mutant *E. coli* strain (J5) possessing the core antigen as its major cell wall component, to patients with gram-negative bacteremia improved survival from 62% in patients given control (low-titer) antiserum to 76% in patients given the high-titer antiserum ($p < 0.04$). Most impressively, in patients with endotoxic shock, administration of the high-titer antiserum as an adjunctive therapeutic modality, along with antimicrobial therapy, necessary fluid resuscitation and other standard support measures, improved survival from 24% to 54% ($p < 0.009$).

Recent studies have investigated the efficacy of large doses of corticosteroids, opiate antagonists such as naloxone, prostaglandin inhibitors, and thyroid releasing hormone as adjunctive therapeutic modalities in endotoxic shock (see Chapters 6 and 7 in this volume and Chapters 7, 9, 10 and 12 in Volume 2).

2. CLINICAL FEATURES OF GRAM-NEGATIVE SEPSIS AND ENDOTOXIC SHOCK

The clinical manifestations of major gram-negative infections culminating in the syndrome of gram-negative rod sepsis with endotoxic shock range widely, from no symptoms whatsoever, or only evanescent fever, to fulminant shock with profound metabolic acidosis, anuria, respiratory failure, and disseminated intravascular coagulation, with death occurring within hours (Table 1).

TABLE 1 *Clinical signs and symptoms and laboratory abnormalities suggesting bacteremic infection*[*]

Clinical	Laboratory
Fever (or rarely, hypothermia)	Leukocytosis ($>$10,000/mm^3) or neutropenia,
Chills or shaking rigors	especially with:
Diaphoresis	'Shift to the left' (many granulocyte
Hypotension $\rightarrow$ frank shock[**]	precursors)
Hyperventilation or frank	Vacuolated neutrophils
respiratory failure[**]	Döhle bodies
Clouded mentation[**]:	Coagulation abnormalities of disseminated
Confusion	intravascular coagulation (DIC)[**]:
Delerium	Hypoprothrombinemia ($>$15 sec)
Obtundation	Hypofibrinogenemia ($<$150 mg/dl)
Nonspecific gastrointestinal	Thrombocytopenia ($<$100,000/mm^3)
symptoms:	Increased titers of fibrinogen degradation
Abdominal pain	products
Vomiting	Microangiopathic changes in erythrocytes
Diarrhea	Hypocomplementemia[**]
Jaundice	Liver function abnormalities:[**]
Oliguria[**]	Hyperbilirubinemia (mainly direct)
Cutaneous lesions:	Mild elevations alkaline phosphatase, and
Petechiae, echymosis, mucosal	to lesser degree, transaminases
bleeding	Hypoxemia[**]
Pustules	Respiratory alkalosis (early) $\rightarrow$ metabolic
Ecthyma gangrenosum	acidosis with lactemia[**] (late)
Osler nodes, Janeway lesions	Azotemia $\rightarrow$ frank renal failure[**]
	Hypophosphatemia

[*] From Maki (1981)
[**] Typically seen in endotoxic shock

With most systemic gram-negative rod infections, the major exceptions being cases originating from intravascular devices or deep intraabdominal abscesses, the signs and symptoms of the local infection giving rise to the bacteremia – e.g., pyelonephritis, peritonitis or pneumonia – usually predominate. Fever, associated with chills or even shaking rigors and with diaphoresis, is almost universal at the outset, and temperatures exceeding 40°C are common (Maki et al 1976; McCabe and Jackson 1962b, 1976; Rector 1981). In a careful study of fever patterns in gram-negative bacteremia, Rector (1981) confirmed the earlier observations of Weil et al (1964) that patients developing endotoxic shock have higher fevers on the average than bacteremic patients who do not develop shock. Hypothermia in gram-negative sepsis is rather rare; it is seen primarily in newborns, in the elderly or in severely compromised patients and is associated with a bad prognosis (Bryant et al 1971; Gleckman and Hibert 1982; McCabe and Jackson 1962b; Weinstein et al 1983b).

Most septic patients hyperventilate because of the effects of bacteremia on the medullary respiratory center. Many also show varying degrees of hypotension (systolic arterial pressure less than 90 mmHg) but, as noted earlier, are not necessarily truly in shock. Shock, usually associated with decreased oxygen consumption, oliguria, metabolic acidosis and lactemia (Nishijima et al 1973; Winslow et al 1973), in most cases clinically reflects a large volume of deep infection (e.g., peritonitis or gram-negative pneumonia), a collection of pus under pressure (e.g., an abscess, pyohydronephrosis, obstructive ascending cholangitis), untreated endovascular infection (e.g., continued infusion of intravenous fluid contaminated by gram-negative bacilli), or simply a severely compromised host. It occurs in about one-third of cases on the average, and when present, invariably connotes greatly increased case-fatality (Fig. 1).

Nonspecific gastrointestinal symptoms such as pain, vomiting, or diarrhea and nonlocalized neurologic symptoms such as confusion, delirium or obtundation, are also commonly encountered in symptomatic gram-negative sepsis and can insidiously divert clinicians' attention away from consideration of serious infection (Maki et al 1976).

Nosocomial gram-negative bacteremias, more often than community-acquired cases, may exhibit cutaneous signs of systemic infection, including pustules or metastatic skin abscesses. Black necrotic ulcers or papules with surrounding erythema (ecthyma gangrenosum) occur most frequently in leukemic and other granulocytopenic patients with *P. aeruginosa* sepsis (Dorff

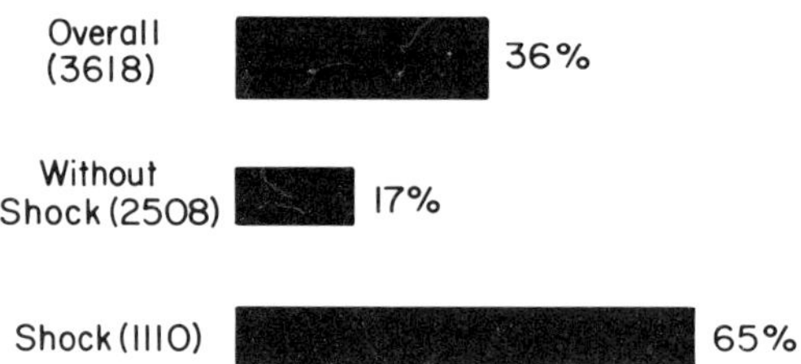

Fig. 1 *Influence of endotoxic shock on case-fatality in 3618 cases of gram-negative bacteremia. Based on data from 10 series reported between 1962 and 1983 (Bryan et al 1983b; Bryant et al 1971; DuPont and Spink 1969; Freid and Vosti 1968; Hodgin and Sanford 1965; Kreger et al 1980b; Maiztegui et al 1965; McCabe and Jackson 1962b; Scheckler 1977; Singer et al 1977).*

et al 1971), but can also be seen with other systemic infections caused by other gram-negative organisms. Blood cultures may be negative, but the presence of ecthyma gangrenosum, a septic endarteritis with cutaneous infarction, should be regarded as diagnostic of systemic infection. Distinguishing features of anaerobic gram-negative bacteremia include jaundice (10–40% of cases), septic thrombophlebitis (5–12% of cases), and distant metastatic suppurative complications (10–28% of cases) (Gorbach and Bartlett 1974). While the origin of septicemia in patients with severe granulocytopenia (less than 500 granulocytes per mm^3) is often obscure, careful repeated examinations may disclose indolent perianal cellulitis, most often caused by *P. aeruginosa* and usually associated with bacteremia (Schimpff et al 1972a), or infection of an intravascular cannula (Maki 1982). Petechiae, ecchymoses, or cutaneous hemorrhaging signal associated coagulation defects, either isolated thrombocytopenia or disseminated intravascular coagulation (DIC) (Beller and Douglas 1973; Cohen and Gardner 1966; Corrigan et al 1968).

Hematologic and biochemical aberrations commonly encountered in gram-negative bacteremic infections are also shown in Table 1. Although leukocytosis, often exceeding 20 000/mm^3, is relatively common (McCabe and Jackson 1962b), morphologic abnormalities of granulocytes in the peripheral blood smear are considerably more specific diagnostically. A marked 'shift to the left' with many immature neutrophils and granulocyte precursors, vacuolization of neutrophils, and intraneutrophilic Döhle bodies strongly suggest microbial invasion of the bloodstream (Emerson et al 1970; Zieve et al 1966). Isolated thrombocytopenia may be the most common coagulation abnormality identified in gram-negative septicemia (Kreger et al 1980b), but hypoprothrombinemia, hypofibrinogenemia, thrombocytopenia, markedly depressed levels of plasminogen and antithrombin III, and high titers of fibrin degradation products can be documented in many patients with proven bacteremia, especially those with septic shock (Büller et al 1982; Corrigan et al 1968; Smith-Ericksen et al 1982). These combined abnormalities almost always denote DIC but rarely pose clinical problems unless refractory shock is present (Corrigan 1977). Hypocomplementemia, particularly of C3, also frequently accompanies septic shock and connotes a high risk of death (McCabe 1973).

Critically ill septic patients, especially those with prolonged shock or with multiple organ failure, are commonly jaundiced in the absence of primary hepatobiliary disease or hemolysis; hyperbilirubinemia, mainly involving the direct fraction, can be marked (higher than 20 mg/dl) and is usually associated with moderate elevation of the serum alkaline phosphatase but with only minimal abnormalities of the transaminases (Banks et al 1982).

Early in the course of bacteremia, almost all patients show respiratory alkalosis because of hyperventilation, but if hypotension is prolonged or if overt shock develops, metabolic acidosis with hyperlactemia rapidly supervenes (Nishijima et al 1973). Hypoxemia due to pulmonary arteriovenous shunting, present in many cases, may culminate in noncardiogenic pulmonary edema (the 'adult respiratory distress syndrome') with severe respiratory failure and high mortality, again, especially if shock cannot be quickly reversed (Towler et al 1983; Kaplan et al 1979; Pepe et al 1982).

Oliguria and azotemia usually accompany shock and have ominous prognostic significance (Wardle 1982).

Systemic signs of bloodstream infection are often masked in neonates, in the elderly and in immunologically compromised patients receiving corticosteroids or other immunosuppressive therapy (Gleckman and Hibert 1982; McCabe and Jackson 1962b; Weil et al 1964). *Most importantly, for an early diagnosis of gram-negative sepsis a high degree of suspicion is indicated, especially in patients with altered host defenses and in a vulnerable setting, in particular those with one or more invasive devices and requiring intensive care unit support.* Fever and mental clouding, especially associated with unexplained tachypnea or hypotension (Table 1), should prompt an immediate and vigorous search for evidence of gram-negative bacillary infection, which should always include blood cultures to identify bacteremia.

3. LABORATORY DIAGNOSIS OF GRAM-NEGATIVE SEPSIS

3.1. Diagnosis of local gram-negative infections

Once the diagnosis of gram-negative sepsis, with or without endotoxic shock, has been considered clinically, the next step diagnostically and one of the most important steps therapeutically is to identify the source of infection – the urinary tract, the lower respiratory tract, the biliary tract or elsewhere within the abdomen or, in a hospitalized patient, an intravascular device – to guide subsequent management. Identifying the initial source of infection may allow surgical extirpation, drainage of an abscess or decompression of an obstructed, infected organ as in obstructive suppurative ascending cholangitis. It also permits selection of more specific antimicrobial therapy, reducing the risk of superinfection and drug toxicity.

In most cases a patient with a clinical picture of gram-negative sepsis shows overt clinical signs and symptoms pointing towards the anatomic site of infection. A rapid diagnostic procedure of considerable use, which unfortunately is greatly neglected in clinical medicine, is the gram-stained smear. In gram-negative bacillary infections of the lower respiratory tract, urinary tract, skin and soft tissues, and with abscesses of all types, a gram-stained smear of tracheal secretions, of urine, or of exudate from the area of purulence will show the infecting microorganism(s) in large numbers in most cases, usually with sheets of polymorphonuclear leukocytes. The gram-stained smear permits presumptive laboratory diagnosis not only of gram-negative infection, but of mixed infections where gram-negative bacilli are co-pathogens with gram-positive cocci or fungi. Anaerobic infections should be strongly suspected by the clinical setting (e.g., peritonitis or an intraabdominal abscess), but a gram-stained smear showing bacteria of diverse morphologies is strongly suggestive of anaerobic polymicrobial infection and should influence selection of anti-infective therapy.

When the clinical setting or the gram-stained smear indicates that anaerobes may well be major pathogens, with or without aerobic gram-negative bacilli, appropriate anaerobic culture techniques should always be performed. This usually requires collection of specimens by appropriate anaerobic techniques, with the use of closed syringes for aspiration of exudate or other materials.

3.2. Laboratory diagnosis of bacteremia

Whereas the patient's clinical picture may strongly suggest bacteremic infection, laboratory confirmation of bacteremia and complete microbiologic characterization of the blood pathogen are imperative for optimal clinical management and, in the hospital, for epidemiologic purposes as well. Laboratory methods for diagnosis of bacteremia by conventional culture methods are thoroughly addressed in several excellent recent monographs (Reller et al 1982; Washington, 1978). Salient aspects of the laboratory diagnosis of bacteremia will be reviewed briefly.

3.2.1. Drawing blood cultures

In general, except for the early phase of brucellosis, typhoid fever, and endovascular infections such as endocarditis or infusion-related sepsis, which produce continuous bacteremia, most gram-negative bacteremias encountered clinically are intermittent in nature (Maki 1981). Thus, except in the aforementioned states, where the timing of blood cultures is probably immaterial, with most infections blood cultures should be drawn at the first clinical indication of sepsis, ideally during a chill or rigor and before the pyrexic episode has begun to abate. In general, the patient's clinical condition will dictate the intervals between cultures; that is, if endotoxin shock is already present or the patient shows signs of acute bacterial endocarditis, blood cultures should be drawn immediately and antimicrobial therapy started without delay.

The fact that a patient may be receiving systemic antimicrobial therapy should never preclude obtaining blood cultures when clinical signs and symptoms suggest bacteremia (Anderson et al 1976). However, in this circumstance, blood cultures obtained immediately before a dose of antibiotics is due to be administered, when the blood antibiotic level is lowest, may produce a higher yield. There is little evidence that adding beta-lactamase to the media materially improves the yield of blood cultures in patients receiving antibiotics; conversely, penicillinase readily becomes contaminated, and outbreaks of gram-negative rod pseudobacteremia have been traced to such contamination (Maki 1980); if penicillinase is to be used, control cultures of (uninoculated) penicillinase-containing media are strongly recommended.

A new commercial system, the Antibiotic Removal Device® (Marion Scientific, Kansas City, MO), incorporates a solid-phase resin in the liquid media which avidly binds antimicrobial drug present and appears to significantly enhance the yield of blood cultures in patients receiving systemic antimicrobial therapy when used in conjunction with conventional blood culture systems (Hansen et al 1983).

Another recent advance in laboratory diagnosis of bacteremia uses immediate lysis of blood cells in the specimen, followed by centrifugation and quantitative culture on solid media (the 'Isolator' system; E.I. du Pont de Nemours and Co., Wilmington, DE). Comparative trials show that the technique permits significantly more rapid detection of bacteremia (and especially, fungemia) and microbiologic identification and susceptibility testing of the blood pathogens but, except for fungi, little improvement in sensitivity in detection of bacteremia (Carey 1984; Kiehn et al 1983; McLaughlin et al 1983).

Up to 25% of all blood cultures in some hospitals are positive with microorganisms subsequently judged to be contaminants (Washington 1978). It is clear that the experience of the person obtaining the blood cultures heavily influences the rate of contamination. Studies have shown that blood cultures obtained by venipuncture teams have two- to five-fold lower rates of contamination than cultures obtained by medical students or house officers (MacGregor and Beaty 1972; Washington 1978).

Before drawing blood cultures, the skin must first be disinfected. Iodine-containing agents such as tincture of iodine (1–2% iodine and 70% alcohol) or an iodophor are most effective (Selwyn and Ellis 1972; White et al 1970), but 70% alcohol is an acceptable alternative if it is left on for at least 30 seconds before performing venipuncture (Lee et al 1967). Quaternary ammonium antiseptics such as aqueous benzalkonium should never be used; these agents are notoriously prone to becoming contaminated by gram-negative bacilli such as *P. aeruginosa* or *Enterobacter* species, and epidemics of both true bacteremia and pseudobacteremia have derived from such contamination (Maki 1981).

It is common practice in many intensive care units to draw blood cultures through central venous or arterial catheters, or in neonates, through umbilical catheters. The several studies that have prospectively evaluated this practice found that blood cultures obtained through catheters in adults show reasonable concordance with cultures drawn by conventional venipuncture, but that rates of false-positive (contaminated) cultures are slightly higher than with percutaneously drawn cultures (Felices et al 1979; Tonnesen et al

1976). In contrast, most studies have shown that blood cultures drawn through umbilical catheters are contaminated up to 45% of the time (Anagnostakis et al 1975; Nelson et al 1965), but a small study by Cowett and coworkers (1976) found that blood cultures drawn through umbilical catheters immediately after they had been inserted did not have a higher rate of false positivity than concomitant cultures obtained by peripheral venipuncture. Recently, prospective studies have shown that the use of quantitative blood cultures drawn through central venous catheters, either hyperalimentation catheters (Snydman et al 1982) or surgically implanted Hickman or Broviac catheters (Raucher et al 1983), can permit noninvasive diagnosis of catheter-related sepsis.

The practice of routinely drawing blood cultures through indwelling vascular catheters ought not to be encouraged because of the risk of introducing contamination into the infusion with added manipulations and the higher frequency of contamination. If, however, to preserve dwindling superficial veins it is considered necessary to use a vascular catheter to obtain blood cultures, we suggest that an attempt be made to use recently inserted catheters and to draw every other specimen by conventional percutaneous venipuncture.

Only closed blood culture systems – plugged bottles entered with a needle – should ever be used clinically (Washington 1978). The specimen may be drawn with a syringe and inoculated directly into the bottle or a transfer set may be used, or the specimen may be drawn directly into an evacuated tube containing the media (the Vacutainer® system, Becton-Dickinson).

Several investigators (Hall et al 1976; Tenney et al 1982; Washington et al 1978; Weinstein et al 1983a) have shown the critical importance of obtaining a sufficient volume of blood for culture, especially in adults, to maximize the yield. In infants, where the concentration of organisms in bacteremia is usually considerably higher than in adults (frequently more than 10^2 organisms/ml; Dietzman et al 1974; Santosham and Moxon 1977), 1- to 3-ml specimens are probably adequate in most cases. In a comparative study of a micro-blood culture technique in neonates, Mangurten and LeBeau (1977) found that a 0.02-ml capillary blood specimen was positive for the same organism isolated from peripheral venous blood cultures in 8 (73%) of 11 bacteremias. Franciosi and Favero (1972) have offered data to suggest that in neonates with sepsis, a single culture will usually be diagnostic. In adults, a considerably larger volume is needed; obtaining at least 20 ml, ideally 30 ml, per drawing, each specimen consisting of 10 or 15 ml inoculated into a

50-ml or 100-ml bottle, significantly improves the yield as compared with obtaining only 5-ml aliquots at each drawing and culturing a smaller total volume (Tenney et al 1982; Washington 1978; Weinstein et al 1983a). There is rarely any need to obtain more than two 15-ml cultures or three 10-ml cultures in any 24-hour period; if at least 30 ml of blood are cultured, 99% of detectable bacteremias should be detected (Fig. 2).

Unless the laboratory uses a 'dual system' (routinely splitting each blood culture specimen and incubating one-half aerobically and the other half anaerobically), half of each specimen should be inoculated into a nonvented (anaerobic) bottle and the remainder into a vented (aerobic) bottle (Washington 1978). Determining whether a bottle giving growth represents bona fide, clinically significant bacteremia as opposed to contamination is greatly aided by performing a separate venipuncture for each set of blood

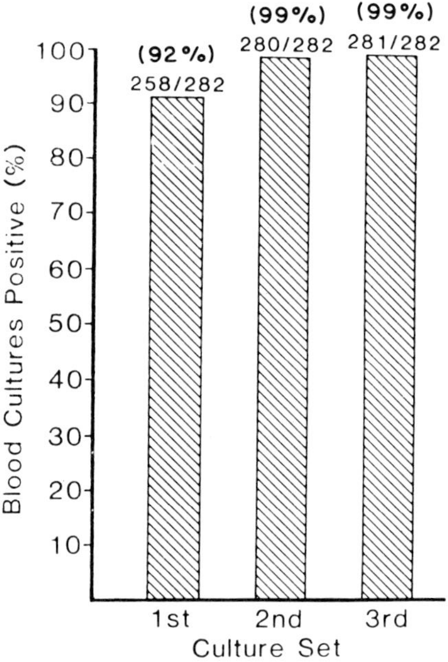

Fig. 2 *Cumulative rates of positivity in three sets of 15 ml blood cultures from patients with proven bacteremia. It is apparent that with 15 ml samples 99% of true culture-documentable bacteremias in adults will be detected by drawing two sets of blood cultures (or sampling at least 30 ml of blood). Reproduced from Weinstein et al (1983a), with permission.*

cultures drawn. Moreover, using a dual (anaerobic-aerobic) system should produce higher yields: between 6% and 8% of bacteremias are caused by obligate anaerobes (Gorbach and Bartlett 1974; Washington 1978). Conversely, recovery of *P. aeruginosa* is considerably enhanced when blood culture bottles are transiently vented (Washington 1978).

Quantitative blood cultures have been evaluated in several studies, using pour-plate techniques (Dietzman et al 1974; Dupont and Spink 1969; Kluge and Dupont 1973; Kreger et al 1980a; Santosham and Moxon 1977). In general, pour-plate cultures proved less sensitive, probably reflecting the smaller volume of blood culture, than conventional techniques but permitted slightly earlier detection of bacteremia. Bacterial counts in bacteremic children, particularly neonates, were found on the average to be considerably higher (50% of cases > 50 organisms/ml) than in adults (80% < 10 organisms/ml); counts exceeding 10^3 organisms/ml in children signalled a 90% likelihood of meningitis. In both children and adults, counts correlated inversely with survival. MacGregor and Beaty (1972) found that quantitative blood cultures were positive in 68% of true bacteremias but in only 6% of false-positive (contaminated) cultures.

3.2.2. *Processing blood cultures*

It is beyond the scope of this chapter to discuss laboratory methods of processing blood cultures and the merits of various blood culture media. This area has also been thoroughly reviewed in several excellent monographs (Reller et al 1982; Washington 1978). Several points, however, are extremely important epidemiologically as regards nosocomial gram-negative infections and their control (Maki 1981):

a. All blood cultures should be identified at least through species, especially gram-negative bacilli. Failure of hospital laboratories to do so in 1970 and 1971 resulted in large outbreaks of infusion-related *Enterobacter cloacae* and *Enterobacter agglomerans* septicemia in hospitals which failed to recognize the outbreaks at the time (Maki et al 1976); their epidemics were identified only in retrospect on reviewing hospital laboratory records of positive blood cultures.

b. Similarly, complete antimicrobial susceptibility testing using standardized methods should be performed routinely on all blood isolates. The susceptibility pattern (antibiogram) is often of considerable value as an epidemiological marker (Aber and Mackel 1981).

c.   An isolate from each clinically relevant bacteremia should be routinely saved for 12–24 months, ideally frozen at −70 °C for future study, should an epidemic of bacteremias be suspected. This is done routinely in most teaching hospitals and has aided immeasurably in the epidemiologic evaluation of many nosocomial outbreaks of gram-negative bacteremia reported in the literature.

3.2.3.   Interpretation of positive blood cultures

Between 6% and 15% of blood cultures in most hospitals will be positive, representing true bacteremia approximately half of the time and contamination introduced through processing the culture in the rest (Washington 1978). Multiple blood cultures positive for one species from a given patient, occurring with a clinical picture consistent with bloodstream infection (Table 1), especially if the same species is also isolated from an identifiable local infection such as the urinary tract, nearly always represents true bacteremia. The increasing frequency of the phenomenon of epidemic pseudobacteremia bears reminding, however (Maki 1980).

Based on two large prospective studies of blood culturing in documented human bacteremic infection by MacGregor and Beaty (1962) and Weinstein et al (1983a), the following microbiologic criteria are offered to help discriminate between true bacteremia and contaminated blood cultures. Isolation of *Streptococcus pneumoniae*, group A streptococci, *H. influenzae* type b, *Neisseria meningitidis*, *Neisseria gonorrhoeae*, *Listeria monocytogenes*, or *Candida*, nearly always denotes true bacteremia, even when recovered from only a single blood culture. Recovery of *Staphylococcus aureus*, enteric gram-negative bacilli, *P. aeruginosa* or anaerobic gram-negative bacilli reflects true bacteremia in over 95% of cases. *Staphylococcus epidermidis* and other coagulase-negative staphylococci, *Bacillus* species, and *Propionibacterium acnes* (diphtheroids) are almost always contaminants unless recovered from multiple blood cultures. Isolates of alpha-hemolytic streptococci, enterococci, *Clostridium* species or more than one organism from a blood culture represent true bacteremia and contamination in approximately equal frequency.

Weinstein and coworkers (1983a) have demonstrated the value of obtaining multiple blood cultures in discerning between true bacteremia and contamination. In their large prospective study of 500 bacteremias, they found that, using 15-ml blood culture specimens for each culture, in patients with endocarditis or other endovascular infections, 95–100% of all cultures drawn were positive. For patients with other bloodstream infections, approximately 75%–80% of all cultures obtained in series were positive. In contrast, with a positive blood culture subsequently considered to represent contamination, the probability that subsequent cultures would be positive was very low, and even if subsequent cultures yielded growth, the organisms recovered were almost always different from those found in the initial culture (e.g., *S. epidermidis* followed by *P. acnes*).

3.2.4.   Noncultural techniques

The gravity of gram-negative bacteremia and other serious systemic bacterial infections and the critical importance therapeutically of making the earliest diagnosis possible have prompted intensive efforts to develop methods for diagnosing bacterial infections more rapidly, by use of noncultural techniques. Anhalt of the Mayo Clinic recently contributed a scholarly review of progress in this field (Anhalt 1978). The use of the Limulus lysate assay for detection of circulating endotoxin is reviewed elsewhere Chapter 4.

4.   SOURCES OF SEPSIS AND MICROORGANISMS ASSOCIATED WITH GRAM-NEGATIVE ENDOTOXIC SHOCK

4.1.   Risk factors for gram-negative sepsis and endotoxic shock

4.1.1.   Pathophysiologic considerations
Bacteremia, the most extreme form of infection, occurs in one of three clinical settings (Maki 1981):

a.   Localized infection no longer can be contained and overwhelms the host defenses of even the healthiest, immunologically competent patient; here the sheer volume of local infection is usually large (e.g., peritonitis) or the infection occurs in association with pus under pressure (e.g., a large abscess, pyohydronephrosis, or suppurative cholangitis).

b.   Bacteremia originates from a seemingly trivial local infection (e.g., a small area of cellulitis) or occurs in the absence of any identifiable local infection whatsoever ('cryptogenic bacteremia'), usually in a neonate or infant, or an immunocompromised patient (e.g., a leukemic).

c. Microorganisms are introduced directly into the bloodstream, bypassing local host defenses, through an intravascular device.

4.1.2. Risk factors for gram-negative bacteremia.

Whereas *S. pneumoniae*, *H. influenzae*, *N. meningitidis*, and *S. aureus* often produce bacteremia without an identifiable local source of infection, gram-negative bacilli, which cause most cases of endotoxic shock, rarely cause bacteremia or shock unless the patient is intrinsically severely immunocompromised or the microorganism has been permitted access to normally inaccessible anatomic sites of the bloodstream by breakdown of anatomic barriers (e.g., by trauma, surgery, or a gastrointestinal perforation) or because of the presence of an invasive device (e.g., a urethral catheter, an endotracheal tube, or an intravascular device). Not surprisingly, as a consequence approximately two-thirds of all gram-negative bacteremias and an even higher proportion of cases of endotoxic shock are nosocomial, that is, they derive from an infection originating in the hospital.

Most of the published studies of gram-negative bacteremia (Bryan et al 1983b; Bryant et al 1971; DuPont and Spink 1969; Freid and Vosti 1968; Haley et al 1981; Hooten et al 1981; Kreger et al 1980b; Lewis and Fekety 1969; Maiztegui et al 1965; Maki 1981; McHenry et al 1975; Myerowitz et al 1971; Rector 1981; Scheckler 1977; Setia and Gross 1977; Weil et al 1964; Weinstein et al 1983b) have reported host conditions and therapeutic factors encountered in a high frequency in patients developing gram-negative bacteremia, including extremes of age, fatal underlying diseases (especially hematologic malignancy or cirrhosis with ascites), biliary tract disease, granulocytopenia, multiple trauma or burns, diabetes mellitus, corticosteroid or other immunosuppressive therapy and a previous local infection (Table 2). Many of these factors are closely related. Unfortunately, except for the rigorous analysis reported by Hooten et al (1981), these studies have not examined the distribution and frequency of putative risk factors in nonbacteremic patients which is necessary for determining the true magnitude of relative risk associated with these factors as regards development of gram-negative bacteremia. Hooten and his CDC coworkers (1981) carried out a prospective, multivariate categorical risk factor analysis in 169 526 patients hospitalized in 338 United States hospitals in 1975–76 to identify factors independently predisposing to nosocomial infection at various sites, including bacteremia.

TABLE 2   *Reported risk factors for gram-negative bacteremia, endotoxic shock, and death due to gram-negative sepsis**

Gram-negative bacteremia**		Endotoxic shock with gram-negative bacteremia	Death due to gram-negative bacteremia
Host factors	Therapeutic factors		
Fatal underlying disease or 'high intrinsic risk'***	*Confinement in critical care unit or newborn nursery*	Fatal underlying disease	Fatal underlying disease
Newborn, especially premature or low birth weight	Systemic antimicrobial therapy	Advanced age	Endotoxic shock
Advanced age (>60 years)	*Invasive vascular device (especially arterial pressure monitoring system)*	Prior corticosteroid therapy	Advanced age
Cirrhosis, especially with ascites, or biliary tract disease	*Receipt of large volumes of parenteral fluids or blood products*	*P. aeruginosa* bacteremia	Granulocytopenia
Diabetes mellitus	*Hemodialysis*	Bacteremia from:	Multiple organ failure, especially renal or respiratory failure
Granulocytopenia	Nonvascular invasive devices:	peritonitis	Congestive heart failure
Multiple trauma or burns	urinary tract catheter or	pyohydronephrosis	Diabetes mellitis
Corticosteroid, other immunosuppressive therapy***	other urinary instrumentation***; endotracheal intubation and mechanical ventilatory support	obstructive cholangitis	Prior antimicrobial or corticosteroid therapy
		pneumonia	Nosocomial origin
Previous local infection***, especially intraabdominal or pneumonia, or any local infection with *Bacteroides*, *Serratia*, *Acinetobacter* or *Providencia*	Major surgical operation***	Inappropriate antimicrobial therapy	Hypothermic response
			Unknown source or respiratory tract source
			P. aeruginosa bacteremia
			Inappropriate antimicrobial therapy

* From published reports on gram-negative bacteremia including risk-factor analyses (Bryan et al 1983b; Bryant et al 1971; Dupont and Spink 1969; Freid and Vosti 1968; Haley et al 1981; Hooten et al 1981; Jarvis et al 1984; Kreger et al 1980b; Lewis and Fekety 1969; Maiztegui et al 1965; McCabe and Jackson 1962b; McHenry et al 1975; Myerowitz et al 1971; Rector et al 1981; Scheckler 1977; Setia and Gross 1977; Weil et al 1964; Weinstein et al 1983b)

** Risk factors for epidemic nosocomial bacteremia in italics (Maki 1981)

*** Independent risk factors for nosocomial gram-negative bacteremia, identified by multivariate categorical data analysis (Hooten et al 1981)

Previous local infection, high 'intrinsic risk,' (a quantitative index incorporating severity of underlying diseases and operative procedures; Hooten et al 1980), urinary catheterization, and corticosteroid or immunosuppressive therapy were found to be strongly associated with the development of nosocomial bacteremia.

As shown in Table 2, the risk of developing gram-negative septicemia, particularly in the hospital, is 2–4 times higher in neonates, especially premature and low-birth infants (Hemming et al 1976), and in elderly patients who are more than 60 years of age. Moreover, patients of any age with a fatal underlying disease, who are as a rule immunocompromised or granulocytopenic because of their disease or its treatment, develop bacteremia 10 to 15 times more frequently than patients without a fatal underlying disease. Patients with leukemia or renal or bone marrow transplants experience very high rates of gram-negative bacteremic infection, ranging up to 500 cases per 1000 (Buckner et al 1978; Murphy et al 1976; Singer et al 1977); most of these bacteremias are hospital-related. While compromised patients are clearly more susceptible to infections from invasive devices of all types, most of their bacteremias fall into the first or second aforementioned categories pathophysiologically.

Patients who have undergone major surgical operations or sustained multiple trauma and who require prolonged intravenous therapy or hemodynamic monitoring and exposure to other invasive devices (Table 2), especially within a critical care unit, also experience very high rates of nosocomial bacteremia, approaching 300 cases per 1000 (Daschner et al 1982; Goldmann et al 1981; Maki 1981; Miller et al 1973; Northey et al 1974), even though these patients in general are not intrinsically immunocompromised and rarely have fatal underlying diseases. The vast majority of these infections are causally related to surgery or exposure to invasive devices such as urinary catheters, endotracheal tubes and mechanical ventilatory support, hemodialysis, or especially, intravascular devices of numerous types.

Critical analyses by McGowan et al (1977) and Spengler and Greenough (1978) suggest that at best only about one-fourth of endemic nosocomial bacteremias can be prevented by more consistent application of known control measures with regard to devices. Developing more effective means to eradicate hospital reservoirs of nosocomial pathogens and prevent, or at least delay, colonization by hospital organisms would seem a prime goal to reduce the incidence of endemic gram-negative bacteremia (Maki 1978).

The extraordinary increase in nosocomial infections in the past three decades, particularly bacteremic infections, caused by gram-negative bacilli, parallels the greatly increased use of systemic antibiotics in hospitals (Finland et al 1959; McGowan et al 1975). Most of these organisms colonize the human intestine, but are not normally found in significant numbers, if at all, in the respiratory tract or on skin. Severe underlying disease or hospitalization alone, with or without exposure to broad spectrum antibiotics, produces marked floral shifts (Harris et al 1976; Johanson et al 1969, 1972; LeFrock et al 1979a; Pollack et al 1972; Rose and Babcock 1975; Rose and Schreier 1968; Schimpff et al 1972b; Stratford et al 1968). Even in patients without major underlying diseases, broad-spectrum antimicrobial therapy alone produces profound floral shifts and gram-negative bacilli become the predominant species colonizing these areas (Goodpasture et al 1977; Johanson et al 1972; Pollack et al 1972; Rose and Schreier 1969; Tillotson and Finland 1969; Weinstein et al 1954). Studies have demonstrated that in granulocytopenic patients, heavy cutaneous or respiratory colonization, or gastrointestinal colonization by hospital-acquired strains of *P. aeruginosa*, precedes invasive infection with these organisms and is a major risk factor for the development of bacteremia (Bodey 1970; Schimpff et al 1972b).

4.1.3. *Risk factors for endotoxic shock*

As noted earlier, most patients with gram-negative sepsis do not develop endotoxic shock. Endotoxic shock is in most cases a consequence of far advanced and overwhelming gram-negative sepsis, in almost all cases associated with bacteremia. A number of investigators have variably examined risk factors (Table 2) for development of endotoxic shock in patients with gram-negative bacteremia (Kreger et al 1980b; McCabe and Jackson 1962b; Rector 1981; Weil et al 1964; Weinstein et al 1983b). Elderly patients and patients with fatal underlying diseases, who are already at high risk of developing gram-negative bacteremia, are also at high risk of developing endotoxic shock, as are patients developing gram-negative sepsis in the setting of prior antimicrobial or corticosteroid therapy.

It has not been established whether specific gram-negative species, with the possible exception of *P. aeruginosa*, are more virulent and likely to produce shock and death with gram-negative sepsis than other species. An analysis of cases from the recent antibiotic era, occurring since the availability of effective antibiotics for treatment of *P. aeruginosa* infections, suggests strongly that *P. aeruginosa* bacteremia is associated with endotoxic shock more frequently than septicemia caused by Enterobacteriaceae, anaerobic gram-negative bacilli, or multiple organisms (Table 3). However, most patients with *P. aeruginosa* sepsis have one or more host risk factors predisposing to endotoxic shock, such as a fatal underlying disease (Kreger et al 1980b), and more sophisticated risk factor analysis, using multivariate techniques, will be required to determine conclusively whether *P. aeruginosa* or other gram-negative species indeed have greater intrinsic virulence as regards production of endotoxic shock. The considerably lower frequency of endotoxic shock encountered in patients with bacteremia caused by gram-negative anaerobes, which this analysis shows (Table 3), is noteworthy and unexplained, but consistent with the findings of studies of microbial virulence in the new Gorbach-Bartlett animal model of intraperitoneal infection (Weinstein et al 1975). Although several investigators (Dupont and Spink 1969; Kreger et al 1980a; Weinstein et al 1983b) have reported that polymicrobial bacteremia is associated with a higher incidence of shock and fatal outcome, the analysis shown in Table 3 does not support this assertion.

Gram-negative infection of certain sites appears to be more likely to result in endotoxic shock, in particular, gram-negative sepsis originating from pneumonia, generalized peritonitis, pyelonephritis with obstruction (pyohydronephrosis), and suppurative ascending cholangitis (due in almost all cases

TABLE 3 *Incidence of septic shock with bacteremia caused by various gram-negative bacilli**

	No. of cases with bacteremia	Cases with shock	
		No.	Rate (%)
Enterobacteriaceae	627	198	31.5
P. aeruginosa	78	33	42.3
Bacteroides spp. and other anaerobic gram-negative bacilli	99	17	17.1
Polymicrobial**	139	50	35.9

* From three recent reports of complete data on the incidence of endotoxic shock associated with bacteremias caused by different gram-negative species (Rector 1981; Singer et al 1977; Weinstein et al 1983b).
** Two or more gram-negative species, with or without gram-positive or fungal pathogens.

to biliary obstruction) – even in patients without fatal underlying diseases or other host factors predisposing to endotoxic shock. With all of these infections, the common denominator is an anatomic setting permitting enormous numbers of gram-negative bacilli access to the bloodstream (pathophysiologic setting *a*, Section 4.1.1.). Survival with endotoxic shock caused by these infections, with the exception of gram-negative pneumonia, is dependent on surgical intervention, correction of the pathology producing peritonitis, decompression of an obstructed ureter or common bile duct, or debridement of massively infected burn wound.

Inadequate antimicrobial therapy of gram-negative sepsis with a drug or drug combination which is not active in vitro against the infecting organism or organisms, greatly increases the likelihood of endotoxic shock and a fatal outcome. Kreger et al (1980b) found that inappropriate antimicrobial therapy increased the likelihood that endotoxic shock would develop in patients with gram-negative bacteremia nearly three-fold, irrespective of the patient's underlying disease, compared with the course in bacteremic patients in which appropriate antimicrobial therapy had been given.

4.2. Sources of gram-negative bacteremia and endotoxic shock

As shown in Table 4, the majority of gram-negative bacteremias develop secondarily as a complication of an identifiable local infection: invasive infections of the urinary tract, intraperitoneal infections that occasionally develop de novo but more often following surgery, suppurative ascending cholangitis, skin and other soft tissue infections which include postoperative surgical wound infections and infected burn wounds. Primary gram-negative bacteremias of unknown origin are relatively infrequent in community-acquired bloodstream infections. In contrast, nearly 25% of endemic gram-negative bacteremias originating in the hospital are primary bacteremias (Maki 1981); most of these primary nosocomial cases derive from intravascular devices of various forms.

4.2.1. Urinary tract infections

The urinary tract is the single anatomic site in man where gram-negative bacilli commonly produce infection in healthy, immunologically competent individuals who do not have underlying diseases predisposing to gram-negative sepsis. Up to 20% of healthy women will experience one or more

TABLE 4 *Sources of bacteremia caused by various gram-negative species**

Origin	Total no. of cases	Proportion (%) of bacteremias originating from site**						Percent of total no. cases
		UTI	RTI	GI	Biliary	Skin/WI	Primary or miscellaneous	
E. coli	302	43.7	5.6	9.3	5.0	3.0	33.4	33.6
Klebsiella spp.	114	23.7	17.5	6.1	5.3	4.4	43.0	12.7
Enterobacter spp.	58	27.6	12.0	5.2	5.2	5.2	44.8	6.4
Pseudomonas spp.	96	25.3	30.2	2.1	4.2	8.3	30.2	10.7
Proteus-Providencia	63	53.9	3.2	1.6	3.2	6.3	31.8	7.0
Serratia	24	25.0	25.0	4.2	0	0	45.8	2.7
Anaerobes	83	12.0	1.2	29.0	0	2.4	55.4	9.2
Other	54	20.4	16.7	14.8	3.7	1.8	42.6	6.0
Polymicrobial	105	23.8	8.6	20.0	0.9	0	46.7	11.7
Total	899	31.7	11.1	10.6	3.7	5.0	37.9	100

* From two recently reported series (Kreger et al 1980a; Weinstein et al 1983a)

** UTI = urinary tract infection; RTI = respiratory tract infection; GI = gastrointestinal or intraperitoneal infection; WI = surgical wound infections. Primary or miscellaneous includes cases originating from intravascular devices, meningitis, and unknown sources.

episodes of symptomatic urinary tract infection during their lifetime (Sanford 1975), and many women are subject to recurrent bouts of symptomatic infection; as many as 10% of elderly men and 20% of elderly women have chronic asymptomatic bacteriuria (Kaye 1980). The vast majority of community-acquired urinary tract infections are caused by sensitive strains of *E. coli*. Fortunately, far less than 1% of all symptomatic urinary tract infections occurring in healthy women culminate in gram-negative bacteremia.

It is not surprising, however, considering the ubiquity of the association of gram-negative bacilli with infections of the urinary tract, that the urinary tract is also the most common site of infection culminating in gram-negative bacteremia (Table 4), particularly gram-negative bacteremia originating in the community; close to one-half of all community-acquired gram-negative bacteremias have a urinary tract origin (Bryan and Reynolds 1984a; Scheckler 1977; Setia and Gross 1977). The vast majority of these cases are caused by *E. coli* or *Proteus* species (Bryan and Reynolds 1984a; Kreger et al 1980a; Scheckler 1977). In the hospital, approximately 20% of all endemic nosocomial gram-negative bacteremias derive from invasive urinary tract infections and in contrast, most nosocomial cases are caused by antibiotic-resistant organisms such as *Klebsiella-Enterobacter* species, *Serratia marcescens*, *Proteus-Providencia* or *P. aeruginosa* (Bryan and Reynolds 1984b; Jarvis et al 1984; Maki 1981).

Prospective studies by Krieger et al (1983) and Bryan and Reynolds (1984b) and review of surveillance data from the CDC's National Nosocomial Infection Surveillance study (Jarvis et al 1984) show that between 2.7% and 3.9% of all nosocomial urinary tract infections are complicated by bacteremia; infections with certain organisms, such as *S. marcescens*, appear more likely to produce bacteremia (Krieger et al 1983).

The vast majority of hospital-acquired urinary tract infections, especially those that are complicated by gram-negative bacteremia, are causally related to the presence of an indwelling urethral catheter or another type of urinary manipulation (Hooten et al 1981; Turck and Stamm 1981); additional risk factors found by Hooten and coworkers (1981) in their multivariate analysis for the occurrence of nosocomial urinary tract infection include advanced age, a high index of 'intrinsic risk', prolonged preoperative hospitalization, and a prolonged duration of operation.

All types of urologic manipulation (Table 5) are associated with substantial risk for producing bacteremia, especially manipulations done in the setting of bacteriuria. In a study of the relationship between urologic instrumentation

TABLE 5   *Factors influencing the initial site of infection and microbial pathogen in gram-negative bacteremia and endotoxic shock*

Originating site	Predisposing factors	Major implicated gram-negative bacterial species
Urinary tract	Obstruction Indwelling urinary catheter Urologic manipulation Renal transplantation Major nonurologic surgery Diabetes	*Escherichia coli* *Klebsiella – Enterobacter* *Proteus – Providencia* *Pseudomonas aeruginosa*
Respiratory tract	Aspiration Nosocomial origin, especially associated with prior surgery Intubation and mechanical ventilatory support Immunosuppressive therapy Granulocytopenia	*Klebsiella – Enterobacter* *Serratia* *P. aeruginosa* *E. coli* *Acinetobacter*
Gastrointestinal Bowel	Perforation Diverticulitis or appendicitis Gastrointestinal surgery Immunosuppressive or antineoplastic therapy Granulocytopenia	*E. coli* *Klebsiella – Enterobacter* *Bacteroides fragilis* *Fusobacterium* spp. *P. aeruginosa* Polymicrobial *Salmonella* spp.
Primary peritonitis	Cirrhosis with ascites Children with nephrotic syndrome Peritoneovenous shunt Peritoneal dialysis	*E. coli* *Klebsiella – Enterobacter* *P. aeruginosa*
Cholangitis	Prior biliary tract obstruction Biliary tract surgery Choledochojejunostomy Hepatic transplantation	*E. coli* *Klebsiella – Enterobacter* *P. aeruginosa* *Fusobacterium* spp. Polymicrobial

TABLE 5 *cont'd*

Gynecological	Postpartum endometritis	*E. coli*
	Cesarean section	*Klebsiella – Enterobacter*
	Therapeutic or criminal abortion	*B. fragilis*
		Fusobacterium spp.
	Gynecologic malignancy	
	Gynecologic surgery	
Skin and soft tissue	Surgical or traumatic wound	*Klebsiella – Enterobacter*
	Full thickness burn	*P. aeruginosa*
	Intravascular devices of all forms	*Proteus* spp.
		B. fragilis
	Immunosuppressive or antineoplastic therapy	Polymicrobial
	Granulocytopenia	
	Decubitus ulcer	
	Diabetes	
	Peripheral vascular disease	
Primary bacteremia		
Unidentified source	Intravascular device	*Klebsiella – Enterobacter*
	Postsurgical	*Serratia*
	Intensive care unit	*P. cepacia* and
	Immunosuppressive or antineoplastic therapy	*P. maltophilia*
		Citrobacter spp.
	Granulocytopenia	*Flavobacter*
	Multiple organ failure	*P. aeruginosa*
	Sickle cell disease and other hemoglobinopathies	*Salmonella*
Infusion therapy	Intravascular catheters	*Klebsiella – Enterobacter*
	Contaminated infusate	*Serratia*
	Critical care unit	*Pseudomonas cepacia* and
	Multiple organ failure	*Pseudomonas maltophilia*
	Granulocytopenia	*Citrobacter* spp.
		Flavobacter

and bacteremia, Sullivan and coworkers (1973) did urine cultures immediately before urologic instrumentation and blood cultures before and immediately after the procedure in 300 hospitalized men, nearly one-third of whom had unrecognized, silent bacteriuria; the presence of pre-existent bacteriuria resulted in bacteremia caused by the same organism in 50% of patients undergoing transurethral resection of the prostate, 33% undergoing cystoscopy or urethral dilation, and 2 of 14 patients undergoing insertion of a urethral catheter; only 12% of patients who had sterile urine at the time of the procedure developed bacteremia ($p < 0.001$). One-half of all of the 60 patients experiencing a procedure-related bacteremia developed endotoxic shock. One of the most fundamental control measures for prevention of nosocomial bacteremia is assuring that urine is sterile at the time of an elective urologic manipulation, especially transurethral prostatic resection.

Additional factors that predispose to bacteremia with urinary tract infection (Table 5) include urinary tract obstruction (Gleckman et al 1982; Lewis and Fekety 1969; Myerowitz et al 1971), renal transplantation (Myerowitz et al 1971), and urinary tract infection occurring in the setting of diabetes (Dupont and Spink 1969; Maiztegui et al 1965). With respect to bacteremic urinary tract infection in diabetics, it is conceivable that the increased risk is related to the high frequency of urinary tract manipulation that occurs in diabetic patients because of neurologic complications of their disease.

Despite the frequency with which bacteremia results from urinary tract infection, it is not as important as a cause of death as other local infections associated with gram-negative bacteremia (Bryan et al 1983b; Bryan and Reynolds 1984a, 1984b; Kreger et al 1980b; Weinstein et al 1983b). This may be due to the frequency with which bacteremic urinary tract infections occur in patients with nonfatal underlying diseases and the relative infrequency of the urinary tract as a cause of bacteremia in patients with ultimately fatal and particularly rapidly fatal underlying diseases (Kreger et al 1980b). The case-fatality for community-acquired bacteremic urinary tract infections is substantially lower (13.7%; Bryan and Reynolds 1984a) than for hospital-acquired bacteremias originating from the urinary tract (30.8%; Bryan and Reynolds 1984b). In Bryan and Reynolds' large, prospective study of bacteremic urinary tract infection (1984a, 1984b), all 43 patients who died as a direct consequence of infection had one or more severe underlying diseases. It seems clear that advanced age and major underlying disease are very important factors associated with mortality due to bacteremic urinary tract infection.

Bacteremic urinary tract infections developing in the community, with the exception of occasional cases occurring in patients with chronic indwelling urethral catheters, are much less frequently associated with urinary instrumentation. Most cases, especially those associated with endotoxic shock, derive from pyelonephritis or, in males, prostatic infection (Bryan and

Reynolds 1984a; Gleckman et al 1982). It is extremely important in all patients with bacteremic urinary tract infection, but especially with infections originating in the community in patients without major underlying diseases, to rule out urinary tract obstruction or perinephric or prostatic abscess as the cause of infection, especially if endotoxic shock is present.

Essential measures for prevention of nosocomial urinary tract infection have recently been published (Center for Disease Control 1981; Garibaldi 1981). Prevention of bacteremic urinary tract infection is predicated upon early diagnosis and treatment of prebacteremic urinary tract infections, assuring that the urine is sterile prior to elective urologic manipulations, and meticulous aseptic technique in catheter care and other urologic manipulations that are associated with a risk of producing bacteriuria.

4.2.2. *Pneumonia*

Local infections of the upper respiratory tract are extremely common, irrespective of the presence or absence of underlying disease, but bacteremia deriving from these infections, with the exception of occasional cases of large parapharyngeal or deep neck abscesses, suppurative otitis media, or bacterial sinusitis, is rare and even when documented, is infrequently caused by gram-negative bacilli other than *H. influenzae* type b in children. On the other hand, bacterial pneumonia is commonly associated with secondary bacteremia, especially pneumonias originating in the hospital where nearly two-thirds of cases are caused by gram-negative bacilli (Jarvis et al 1984); community-acquired pneumonia is rarely caused by gram-negative bacilli, even in alcoholic and immunocompromised patients (MacFarlane et al 1982; Phair et al 1983; Sullivan et al 1972). Bacteremic gram-negative pneumonia has a very high mortality, of more than 60% (Bryan and Reynolds 1984c). Thus, bacteremic gram-negative pneumonia is primarily a nosocomial phenomenon, influenced greatly by the profound effects of severe underlying disease (Johanson et al 1969; Schimpff et al 1972; Stratford et al 1968), hospitalization (Harris et al 1976; Johanson et al 1972; LeFrock et al 1979a; Pollack et al 1972; Rose and Babcock 1975; Rose and Schreier 1968), and systemic antimicrobial therapy (Goodpasture et al 1977; Johanson et al 1972; Pollack et al 1972; Rose and Schreier 1968; Tillotson and Finland 1969; Weinstein et al 1954) on the oropharyngeal flora, the source of organisms causing most nosocomial pneumonias. Whereas healthy individuals are only infrequently colonized by gram-negative bacilli, and when colonized are colonized with only small numbers of organisms, gram-negative bacilli are the major aerobic species in the oropharynx of 95% of patients with serious underlying diseases or who have received broad-spectrum antimicrobial therapy (Johanson et al 1969; Pollack et al 1972; Stratford et al 1968). Johanson et al (1972) found in a large prospective study of nosocomial pneumonia occurring in patients admitted to a critical care unit that severe underlying disease, pre-existent pulmonary disease, prolonged hospitalization, antibiotic therapy, intubation and renal insufficiency were significantly associated with upper airway colonization by aerobic gram-negative bacilli. Whereas many of these organisms are acquired from as yet poorly defined extrinsic sources within the hospital, studies by LeFrock et al (1979a) and Schwartz et al (1978) suggest that they may also commonly derive from the patient's own fecal flora.

Risk factors predisposing to nosocomial pneumonia with or without bacteremia (Table 5), particularly that caused by gram-negative bacilli, apart from prior antimicrobial therapy or serious underlying diseases or a high CDC index of 'intrinsic risk' include advanced age, smoking, chronic obstructive pulmonary disease, major surgery – especially extensive thoracoabdominal or abdominal operations – a prolonged duration of operation, confinement in a critical care unit, intubation or tracheostomy with mechanical ventilatory support, and corticosteroid and other immunosuppressive therapy (Garibaldi et al 1981; Hooten et al 1981; Johanson et al 1972; Stevens et al 1974). Risk analyses for development of bacteremia, once gram-negative pneumonia has occurred, are not available, but approximately 5.2% of patients with nosocomial pneumonia develop secondary bacteremia (Jarvis et al 1984); patients with granulocytopenia experience very high rates of bacteremic gram-negative pneumonia (Schimpff et al 1972b).

The organisms most commonly implicated in nosocomial gram-negative pneumonia, with or without bacteremia, are *Klebsiella-Enterobacter* species, *S. marcescens*, *P. aeruginosa*, *E. coli*, and *Acinetobacter* (Tables 4 and 5). Anaerobic gram-negative bacilli such as *Bacteroides* species, the most frequent pathogens in aspiration pneumonias originating in the community (Lorber and Swenson 1974), are rarely associated with bacteremic pneumonia. In contrast, aspiration pneumonias occurring in the hospital are caused almost exclusively by aerobic gram-negative bacilli and *S. aureus* (Lorber and Swenson 1974); many of these pneumonias culminate in bacteremia.

Essential control measures for prevention of bacteremic gram-negative pneumonia include efforts to diagnose pneumonia as early as possible, before it progresses to bacteremia, and the detailed measures for prevention of hospital-acquired pneumonia, which have recently been summarized (Centers for Disease Control 1982a; Veazey 1981). These include recommendations to prevent aspiration and help mobilize secretions postoperatively, and the numerous precautions with respect to aseptic technique in the care of patients with endotracheal or tracheostomy tubes, delivery of mechanical ventilatory support, and care of respiratory therapy equipment.

4.2.3. Gastrointestinal infections

Between 10% and 20% of all gram-negative bacteremias originate from intraabdominal foci of infection (Table 4), most commonly intraperitoneal infections deriving from a gastrointestinal perforation, diverticulitis or appendicitis; primary peritonitis; or suppurative ascending cholangitis (Table 5). The same risk factors predisposing to postsurgical intraabdominal abscesses also predispose to infection of the postoperative surgical wound and are discussed below. With respect to intraperitoneal infection, besides advanced age and serious underlying diseases, immunosuppressive or antineoplastic therapy and granulocytopenia greatly increase the risk of these infections and associated gram-negative bacteremia.

Suppurative cholangitis occurs almost exclusively in the setting of acute biliary tract obstruction, usually by stones or a stricture rather than neoplasm, and if the obstruction is not relieved surgically, fulminant endotoxic shock and death can rapidly occur, even in young and immunologically competent patients (Andrew and Johnson 1970; Welch and Donaldson 1976). Cholangitis with bacteremia can occasionally occur in the absence of identified obstruction with an established functional choledochojejunostomy or following hepatic transplantation (Schroter et al 1976).

Aerobic gram-negative bacilli, particularly *E. coli*, but also *Klebsiella-Enterobacter* and *P. aeruginosa*, especially in nosocomial cases, and enterococcus, *Clostridium*, and *Fusobacterium* species are the major pathogens in ascending cholangitis with bacteremia; in nonbiliary intraabdominal infections – peritonitis or intraperitoneal abscesses – these organisms are joined by *Bacteroides fragilis* and anaerobic streptococci as intraperitoneal and bacteremic pathogens (Tables 4 and 5). The presence of cryptogenic polymicrobial bacteremia usually denotes an intraabdominal source of infection (Hermans and Washington 1970; Ing et al 1981; Lufkin et al 1966).

Primary peritonitis is a well-defined syndrome where aerobic gram-negative bacilli, primarily *E. coli* or *Klebsiella-Enterobacter* and occasionally *P. aeruginosa*, produce unimicrobial peritonitis, associated with bacteremia more than half of the time, without identifiable disruption of the integrity of the gastrointestinal tract or cholangitis (Table 5). This syndrome is seen only very rarely in patients without pre-existent ascites; most patients have advanced cirrhosis (Conn and Fessel 1971; Weinstein et al 1978), although it also occurs in children, especially girls, with nephrotic syndrome (Rubin et al 1975; Wilfert and Katz 1969).

Peritonitis is a very frequent complication of peritoneal dialysis, where *E. coli, Klebsiella-Enterobacter* species, and *P. aeruginosa* are the most common gram-negative pathogens (Kolmos and Andersen 1979; Kraus and Spector 1983; Rubin et al 1980; Vas 1983).

Salmonella is a well-documented cause of both community-acquired bacteremia and also of nosocomial bacteremia (Baine et al 1973; Centers for Disease Control 1974), almost always deriving from symptomatic gastroenteritis. However, in neonates, patients with lymphoma, other severely immunocompromised patients, or postsurgical patients – especially postgastric operations – salmonella can produce primary bacteremia without signs of overt gastrointestinal infection (Centers for Disease Control 1970; Han et al 1967; Lang et al 1967; Rhame et al 1973; Wolfe et al 1971). Occasionally, cryptogenic salmonella bacteremia is the first sign of an occult mycotic (infective) aneurysm (Parsons et al 1983).

4.2.4. *Gynecologic infections*

Gram-negative bacteremia is a rare complication of pelvic infection in the absence of several well-defined risk factors (Table 5): postpartum endometritis, intrapartum fetal monitoring or cesarean section, therapeutic or criminal abortion, gynecologic malignancy, or recent gynecologic surgery (Bryan et al 1984c; Cavanagh et al 1982; Ledger et al 1975; Ledger 1984). Shock or death is rare (Bryan et al 1984c; McHenry et al 1975), probably because most of the infected patients are relatively young and do not have serious underlying diseases, but also because the pelvis is rapidly and easily amenable to surgical therapeutic intervention (Cavanagh et al 1982; Ledger et al 1975). These infections are caused by the same gram-negative species that cause intraperitoneal infections elsewhere in the abdomen, aerobic and anaerobic gram-negative bacilli, primarily *E. coli, Klebsiella-Enterobacter, B. fragilis,* and *Fusobacterium* species (Tables 4 and 5).

4.2.5. Skin and soft tissue infections and postoperative surgical wound infections

Like infections of the upper respiratory tract, bacterial skin and soft tissue infections are exceedingly common in healthy persons but rarely result in bacteremia; moreover, in immunocompetent persons, these infections are caused almost exclusively by *S. aureus* and beta-hemolytic streptococci. However, patients undergoing surgical operations are at significant risk of developing bacterial infection of the postoperative surgical wound, with the degree of risk and microbial profile of infection heavily influenced by the type of operation performed: rates of infection are low (1–2%) and caused primarily by staphylococci in 'clean' operations – elective orthopedic, neurosurgical or cardiovascular procedures – whereas the rate of infection is considerably higher in operations on the head or neck, lower respiratory tract, gastrointestinal tract or gynecologic organs, where aerobic gram-negative bacilli and anaerobic bacteria are the predominant pathogens (Cruse and Foord 1980; National Research Council 1964). Approximately 5.2% of all postoperative surgical wound infections are complicated by secondary nosocomial bacteremia (Jarvis et al 1984).

Risk factors predisposing to surgical wound infection, besides the type of operation performed, include: a prolonged period of hospitalization prior to surgery; a prolonged operation; advanced age; severe underlying disease (or high CDC index of 'intrinsic risk'); malnutrition; corticosteroid or immunosuppressive therapy; and pre-existent other active infections (Cruse and Foord 1980; Edwards 1976; Hooten et al 1981; National Research Council 1964). Control measures for prevention of postoperative wound infection, beyond optimizing the patient's overall medical condition and eradicating active foci of infection prior to surgery, include minimizing the length of preoperative hospitalization, the detailed ritual of surgical asepsis in the operating room, and the judicious use of perioperative surgical antimicrobial prophylaxis, which should always be begun preoperatively (Aber and Garner 1981; Centers for Disease Control 1982b).

Major full-thickness burn wounds invariably become colonized and most develop infection at some stage, usually with gram-negative bacilli, particularly *P. aeruginosa*. Burn wound sepsis is commonly accompanied by gram-negative bacteremia, with or without endotoxic shock (Pruitt 1979). Systemic antibiotics used prophylactically do not prevent burn wound sepsis, but increase colonization by multiple drug-resistant bacteria and fungi. In contrast, topical antimicrobials such as sulfamylon and silver sulfadiazine clearly reduce the incidence of burn wound sepsis (Lindberg et al 1968; Moncrief 1978). Infected burn wounds should ideally be debrided before full-blown bacteremic sepsis occurs, but debridement can itself cause bacteremia (Sasaki et al 1979).

Certain underlying conditions are well known to predispose to 'complicated' cellulitis and other soft tissue infections (Table 5). Granulocytopenic patients are almost uniquely prone to develop bacteremic gram-negative cellulitis, especially of the perianal regions, caused by aerobic gram-negative bacilli, particularly *P. aeruginosa* (Schimpff et al 1972).

Patients who have undergone recent surgery or sustained major soft tissue injuries, with decubitis ulcers (Bryan et al 1983a; Galpin et al 1976), or who are on immunosuppressive drugs, have diabetes or peripheral vascular disease, are prone to develop highly invasive, polymicrobial soft tissue infections termed 'necrotizing fasciitis' or 'synergistic gangrene' (Burbrick and Hitchcock 1979; Casali et al 1980; Rea and Wyrick 1970; Stone and Martin 1972), usually involving the lower extremities. These infections are caused by aerobic gram-negative bacilli and anaerobic organisms, basically by the same pathogens implicated in intraperitoneal infections; bacteremia occurs in up to 40% of cases. Any local soft tissue inflammation occurring in patients at risk for these complex infections must be vigorously pursued diagnostically, at the minimum with microbiologic examination of percutaneous aspirates or biopsies (Stamenkovic and Lew 1984). If skin necrosis is present or gas is apparent in the deep tissues, exploratory surgery and extensive debridement may be in order (Baxter 1972; Stone and Martin 1972). The case-fatality of these complex skin and soft tissue infections that occur almost exclusively in compromised patients is exceedingly high (over 50%), unless aggressive surgical management is employed early in the course.

4.2.6. *Primary bacteremias*

Whereas *S. aureus* and meningococci commonly cause primary community-acquired bacteremia in healthy immunologically competent patients, community-acquired gram-negative bacteremia is rarely primary; a local source, most often the urinary tract but occasionally cholangitis or another cryptic intraabdominal source of infection, can almost always be found if vigorously sought.

In contrast, 20% to 30% of endemic hospital-acquired gram-negative bacteremias are primary bacteremias, and no clear-cut source is identified except for the 5–10% of cases overall that originate from an intravascular device or a contaminated infusion (Maki 1981). It is likely that most primary nosocomial bacteremias are infusion-related, but the culpable cannula or contaminated infusate is often not cultured, and then the source remains unidentified. In a patient receiving infusion therapy of any form, the occurrence of bacteremia without an apparent anatomic site of infection must raise strong suspicion of infusion-related sepsis, and the infusion should be discontinued and appropriately cultured (Maki 1982).

Granulocytopenic patients experience a very high rate of primary gramnegative bacteremia, in many cases unrelated to infusion therapy (Buckner et al 1978; Schimpff et al 1972b; Singer et al 1977). It is hypothesized, but it has not been proved, that many of these cases originate from the gastrointestinal tract. Measures for prevention of infection in the granulocytopenic patient have been published (Pizzo 1984).

Unfortunately, infusion-related sepsis is probably the least frequently recognized nosocomial infection in most hospitals (Maki et al 1976; Maki 1982). Lack of awareness of the problem is probably due to its relative infrequency; the percentage of infusions producing sepsis is so low that any single physician or nurse is unlikely to encounter more than an occasional case. Most endemic infusion-related bacteremias derive from infection of the intravascular cannula (Band and Maki 1979) and are caused primarily by staphylococci or *Candida*, occasionally by gram-negative bacilli (Table 5); contamination of infusate administered through intravascular devices can also produce septicemia, almost always caused by gram-negative bacilli, particularly *Enterobacter* species, *Flavobacterium*, or *Pseudomonas* species other than *P. aeruginosa* (Maki 1982). The vast majority of nosocomial bacteremias traced to contaminated infusate have occurred in an epidemic setting (Maki 1981), but intravenous fluids can occasionally become contaminated during use in the hospital because of the many manipulations of the infusion and then they may give rise to sporadic endemic bacteremias (Band and Maki 1979). The strong association of gram-negative bacilli, particularly members of the tribe Klebsielleae (*Klebsiella, Enterobacter* and *Serratia* spp.), with sepsis traced to contaminated infusate is a consequence of the unique ability of these gram-negative bacilli, compared with other bacteria and fungi, to multiply rapidly in commercial glucose-containing fluids which are used universally for infusion therapy (Maki and Martin 1975). Infusions used for intraarterial pressure monitoring are particularly prone to become contaminated by gram-negative bacilli (Maki and Hassemer 1981), and numerous outbreaks have derived from contamination of multiple patients' intraarterial infusions in an intensive care unit (Maki 1982).

Detailed guidelines for the prevention of infusion-related bacteremia have been published (Centers for Disease Control 1981; Maki 1982); meticulous asepsis during insertion of the intravascular device and in all handling of the infusion, and limiting the duration of placement of peripheral venous and arterial catheters will greatly reduce the risk of related infection.

There is also a significant risk of device-related bacteremia associated with hemodialysis (Dobkin et al 1978; Keane et al 1977). These infections are of two types: those derived from infection of the site used for vascular access and those due to contamination of solutions used in the hemodialysis machine. Relatively few bacteremias originating from the vascular access sites are caused by gram-negative bacilli. In contrast, contaminated dialysate is a very common cause of sporadic or even epidemic pyrogenic reactions or primary gram-negative bacteremias, most often with *P. aeruginosa* and other aerobic 'water organisms' (Maki 1981). Excellent guidelines for prevention of gram-negative bacteremia from contaminated dialysate in hemodialysis have been published (Favero and Peterson 1977).

4.3. Microbiology of gram-negative bacteremia

Over the past three decades, gram-negative bacilli have vastly overshadowed gram-positive organisms as causative agents of bacteremia, especially bacteremias acquired in the hospital. To provide a large representative base of data on both community-acquired and nosocomial cases occurring in American hospitals, microbiologic data on 631 gram-negative bacteremias reported since 1977 from three centers in the United States are displayed in Figure 3.

E. coli remains preeminent, both in community-acquired and nosocomial bacteremias, but accounts for a higher proportion (42%) of cases of gram-negative bacteremia originating in the community. Bacteremias caused by *H. influenzae* type b, *N. meningitidis*, and *N. gonorrhoeae*, and *Salmonella* species are seen almost exclusively in community-acquired cases. Bacteremias caused by *Bacteroides* species are common in both community-acquired and nosocomial bacteremias. In contrast, bacteremias caused by *Klebsiella*, *P. aeruginosa*, and other *Pseudomonas* species, *Serratia*, *Acinetobacter*, *Flavobacterium*, and other 'water organisms' are infrequent in community-acquired cases, but account for over one-half of all endemic

92

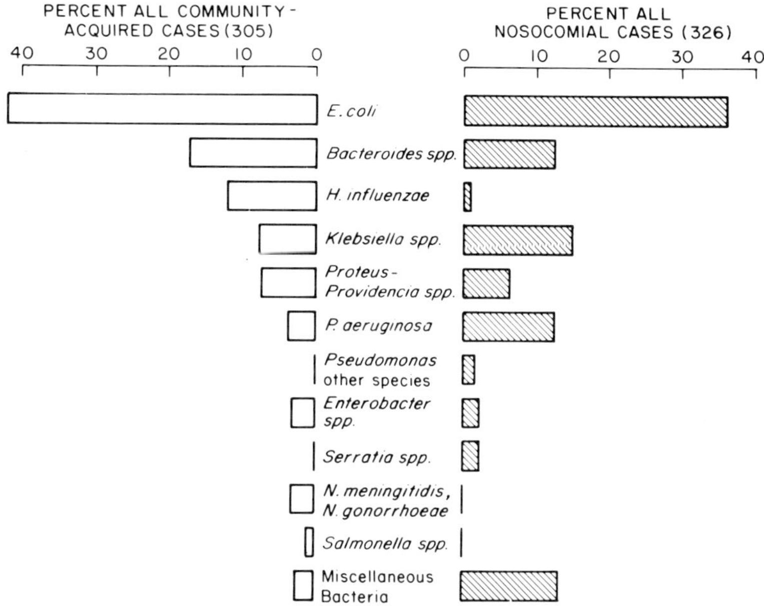

Fig. 3   *Contrast in the microbial profile of 305 community-acquired and 326 nosocomial gram-negative bacteremias. Based on data from three published series encompassing 631 cases (McGowan et al 1976; Scheckler 1977; Weinstein et al 1983a).*

nosocomial gram-negative cases. In the past decade, the number of nosocomial bacteremias caused by *Enterobacter* species, *S. marcescens*, and *P. aeruginosa* has greatly increased (Maki 1981).

Hermans and Washington (1970) reported that at the Mayo Clinic, polymicrobial bacteremia occurs in 6% of cases of bona fide bacteremia and most commonly originates from an intraabdominal infection. Other studies from teaching hospitals have shown 6% to 8% of clinical bacteremias are polymicrobial (Ing et al 1981; Washington 1978).

The likelihood of bloodstream invasion with local infections caused by *Bacteroides* species, *Serratia*, *Acinetobacter*, and *Providencia* is more than 10%, considerably greater than the incidence of secondary bacteremia associated with local infections caused by other gram-negative bacilli (Jarvis et al 1984). The organisms causing nosocomial gram-negative bacteremia are as a rule more resistant to antibiotics than organisms encountered in community-acquired cases, even with cases caused by *E. coli* (McGowan et al 1974).

4.4. Epidemic nosocomial bacteremias

In the past 20 years, there have been numerous reported outbreaks of nosocomial infection, in part reflecting greatly increased awareness of the problem of nosocomial infection and the existence of active surveillance programs in most U.S. hospitals, but also reflecting an increasingly susceptible population of patients being hospitalized. Between 1965 and 1978, 97 epidemics of nosocomial bacteremia were reported in the world literature (Maki 1981) (Fig. 4). Over three-fourths of these outbreaks were caused by aerobic gram-negative bacilli, particularly *Klebsiella* and *Enterobacter* species, *S. marcescens*, *P. aeruginosa*, and especially, other pseudomonas such as *Pseudomonas cepacia* (Fig. 5). *P. cepacia*, in particular, has become a preeminent pathogen in epidemics of nosocomial gram-negative bacteremia, accounting for 14 outbreaks over this period, more than with any other organisms; every outbreak stemmed from infusion therapy or hemodialysis.

In only 19% of the 97 outbreaks, nearly all in newborns, did the anatomic portal of bacteremia and epidemiologic mechanisms elude detection. Outbreaks in hemodialysis units accounted for 7% of the epidemics and miscellaneous primary bacteremias for another 7%. In 11 outbreaks, the putative epidemic infections were ultimately shown to be 'pseudobacteremias' caused by contamination of blood cultures during drawing or during processing in the laboratory (Maki 1980). Outbreaks of local infection at one site giving rise to secondary bacteremias accounted for only 21 (22%) of the 97 epidemics, a striking contrast with the sources of endemic nosocomial bacteremia, most of which are secondary bacteremias and derive from identified local infections (Fig. 6).

Two-thirds of the outbreaks occurred in a closed patient population, usually adults in an intensive care unit (24 epidemics) or hemodialysis center (7 outbreaks), or newborns in a nursery or intensive care unit (26 outbreaks). Moreover, except for epidemics involving neonates, most patients affected in the 86 outbreaks of true bacteremia were not immunologically compromised. Whereas certain host states greatly increase the risk of serious nosocomial infections of all types including bacteremia, epidemics of gram-negative bacteremia occur in large measure irrespective of the character of the patient population and are related primarily to what measures have been taken after admission to the hospital: segregation in a special care unit or exposure to infusion therapy, hemodialysis, or other invasive procedures involving the bloodstream.

There were a dozen outbreaks of acute pyrogenic reactions reported from hospitals in which bacteremia could not be documented. Despite severe illness in many patients, including a picture of endotoxic shock, there were no associated deaths in the twelve reported outbreaks. In each outbreak, all

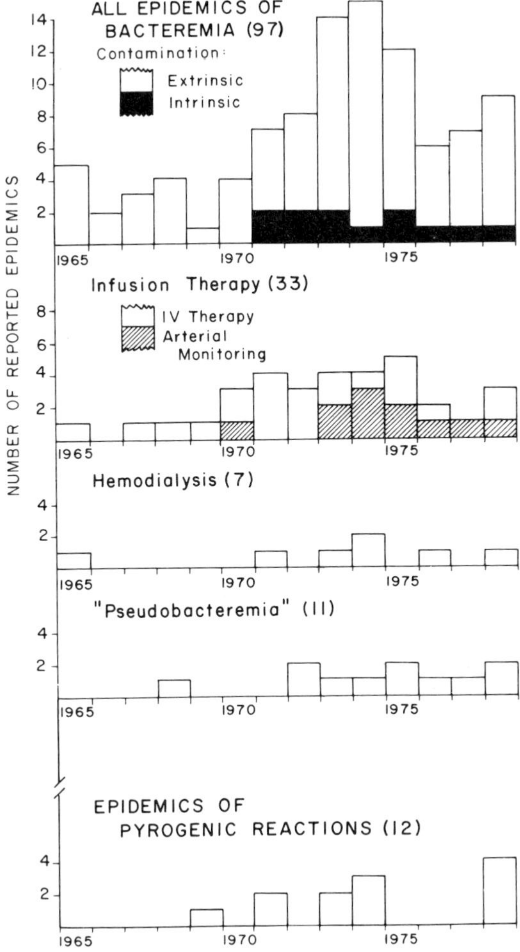

Fig. 4   *Epidemics of nosocomial bacteremia and pyrogenic reactions each year, 1965–1978. Based on data from published reports in English world literature; for epidemics of bacteremia, bacteremias constituted at least 25% of epidemic infections. Reproduced from Maki (1981), with permission.*

95

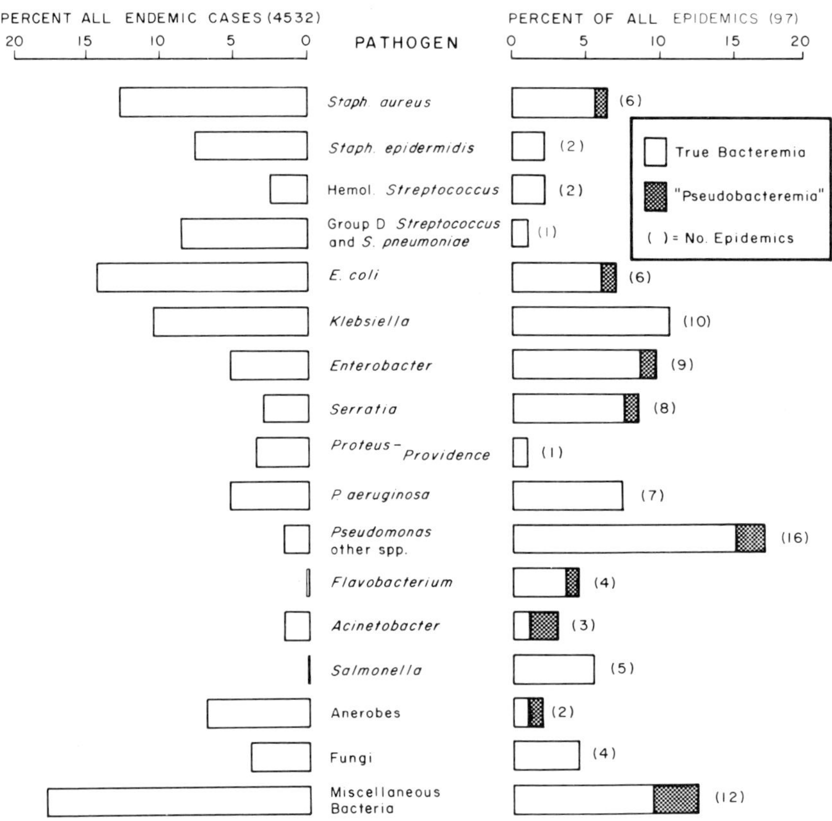

Fig. 5 *Microbial pathogens associated with endemic nosocomial bacteremias and responsible for 97 epidemics occurring between 1965 and 1978. Epidemic pathogens in 'miscellaneous bacteria' include: Citrobacter (4 epidemics); Listeria (2); Bacillus spp. (2); M. chelonei (1); N. meningitidis (1); H. influenzae (1), Moraxella (1). Based on data on 4532 endemic bacteremias from 83 U.S. hospitals which submitted data on 46 821 nosocomial infections occurring in 1976 to the National Nosocomial Infections Study of the CDC. Data on epidemics were compiled from published reports in the English world literature in which bacteremias constituted at least 25% of the epidemic infections. Reproduced from Maki (1981), with permission.*

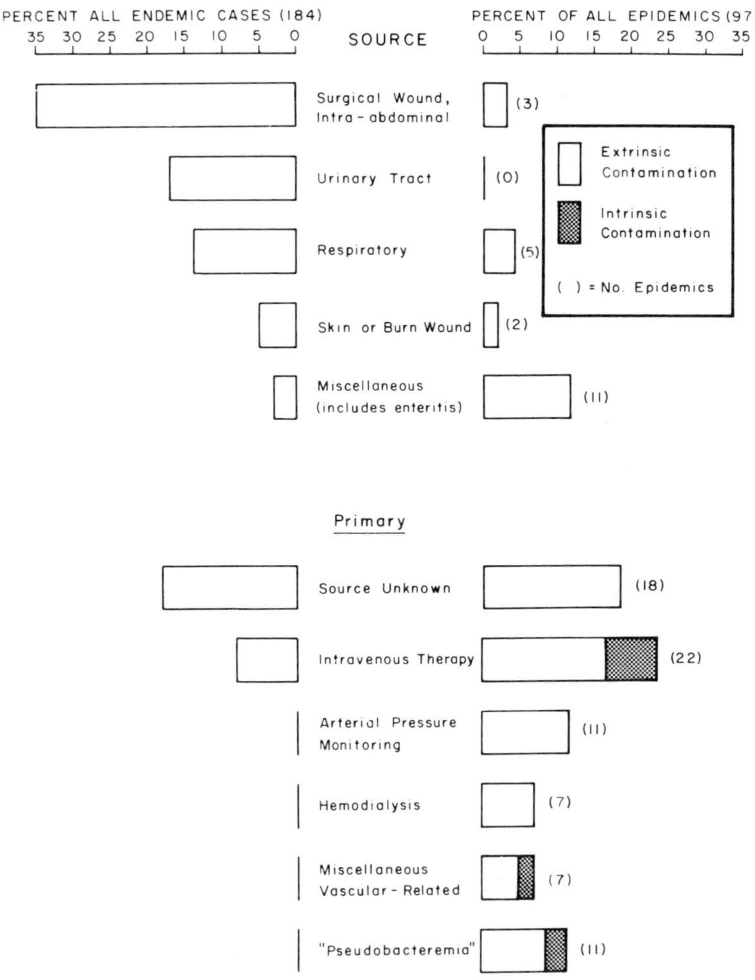

Fig. 6 *Sources of nosocomial bacteremia. Data on 1984 endemic cases from published experiences in 3 United States hospitals: a community hospital, St. Mary's, Madison, WI (Scheckler 1977); a university hospital, Johns Hopkins, Baltimore, MD (Spengler and Greenough 1978); and a municipal hospital, Grady Memorial, Atlanta, GA (McGowan et al 1977). Data on 97 epidemics occurring between 1961 and 1978 compiled from reports in English world literature in which bacteremias constituted 25% or more of the epidemic infections. Reproduced from Maki (1981), with permission.*

affected patients had been exposed to a procedure such as hemodialysis or cardiac catheterization, or had received a common blood product, such as normal serum albumin or plasma protein fraction. Investigations showed that the procedure or the product administered introduced nonviable pyrogens directly into patients' bloodstreams, accounting for a clinical picture indiscernable from gram-negative sepsis. Studies using the Limulus lysate assay suggest strongly that in these outbreaks the pyrogenic substance was gram-negative lipopolysaccharide, that is, endotoxin (Bauer et al 1979; Steere et al 1978).

4.4.1. Epidemics of national scope

Because we live in an era of unprecedented communication, manufactured products can be in thousands of hospitals within days, and contaminated products can be distributed on a vast scale. This culminated in six outbreaks of nosocomial bacteremia involving a number of hospitals, all of national scope, since 1970 (Maki 1981). All intrahospital outbreaks of gram-negative bacteremia stemmed from manufacturer-related (intrinsic) contamination of nationally distributed products, which can be exceedingly difficult to recognize early if the infections are of low frequency or occur in widely scattered hospitals. Intrahospital surveillance networks, such as the CDC's National Nosocomial Infection Surveillance study which carries out surveillance of nosocomial infections in a large cross-section of American hospitals, may be the most effective means of detecting these common-source epidemics early (Maki et al 1976; Maki 1982).

Review of data submitted to this program between January 1970 and 1971 showed that U.S. hospitals that used infusion products of one company experienced a highly significant increase in the incidence of *Enterobacter* bacteremia beginning in June 1970, even though few of the hospitals individually were aware of the existence or the magnitude of this nationwide outbreak before March 1971 (Goldmann et al 1978; Maki et al 1976). No increase in *Enterobacter* bacteremias occurred in American hospitals that used infusion products of other companies.

All four U.S. outbreaks of national scope stemmed from intrinsic contamination of a parenteral product. Any data suggesting iatrogenic contamination of a commercial product, especially if it may have produced illness in man or simply has a potential to do so, should be immediately reported to state and federal public health authorities, and remaining samples of the product should be quarantined and saved for analysis (Maki 1981).

5. MAGNITUDE AND SECULAR TRENDS OF GRAM-NEGATIVE BACTEREMIA

5.1. Bacteremias occurring endemically

Bacteremic infections have long been a subject of great interest, both clinically and investigatively, which is largely due to their clinical gravity and also to an almost mystical aura surrounding infection of the human bloodstream. Over 100 papers have been published in the past 20 years alone reporting morbidity and case-fatality rates associated with bacteremic infections in individual institutions.

The most comprehensive studies longitudinally are those of all types of bacteremia by Finland and his coworkers (McGowan et al 1975) at the Boston City Hospital covering the period 1935-1972 (Fig. 7) and those of gram-negative bacteremia by Dupont and Spink (1969) at the University of Minnesota Hospitals and by McCabe and his coworkers at the University of Illinois and Boston University Medical Center (Kreger et al 1980a, 1980b; McCabe and Jackson 1962a, 1962b). From the studies presented in Table 6 it is clear that the incidence of gram-negative rod bacteremia has risen progressively over the past four decades (Fig. 8). Although rates of bacteremia vary considerably from hospital to hospital, it can be seen in Table 6 that between one and ten patients per 1000 hospital admissions developed a gram-negative bacteremia, about two-thirds of which are hospital-acquired. Rates are lowest in community hospitals (approximately 1–4 cases per 1000) and considerably higher in university and municipal institutions (6–10 cases per 1000).

There has been considerable controversy as to the true magnitude of the problem with gram-negative bacteremia on a national scale (McCabe 1974; Wolff and Bennett 1974). The most recent and probably most representative projections, based on data of 1.3 million patients hospitalized in the 83 hospitals submitting data to the National Nosocomial Infection Surveillance Study in 1976, suggest that on the average there are approximately five nosocomial bacteremias per 1000 patients hospitalized in American hospitals (Dixon 1978), two-thirds of which are caused by gram-negative bacilli. Considering that there are nearly as many community-acquired gram-negative bacteremias as nosocomial cases and by applying the estimated rate of five gram-negative bacteremias per 1000 rate to the approximately 40 million persons hospitalized in the United States each year, it can be estimated that there are upwards of 200 000 gram-negative bacteremias in this country each year.

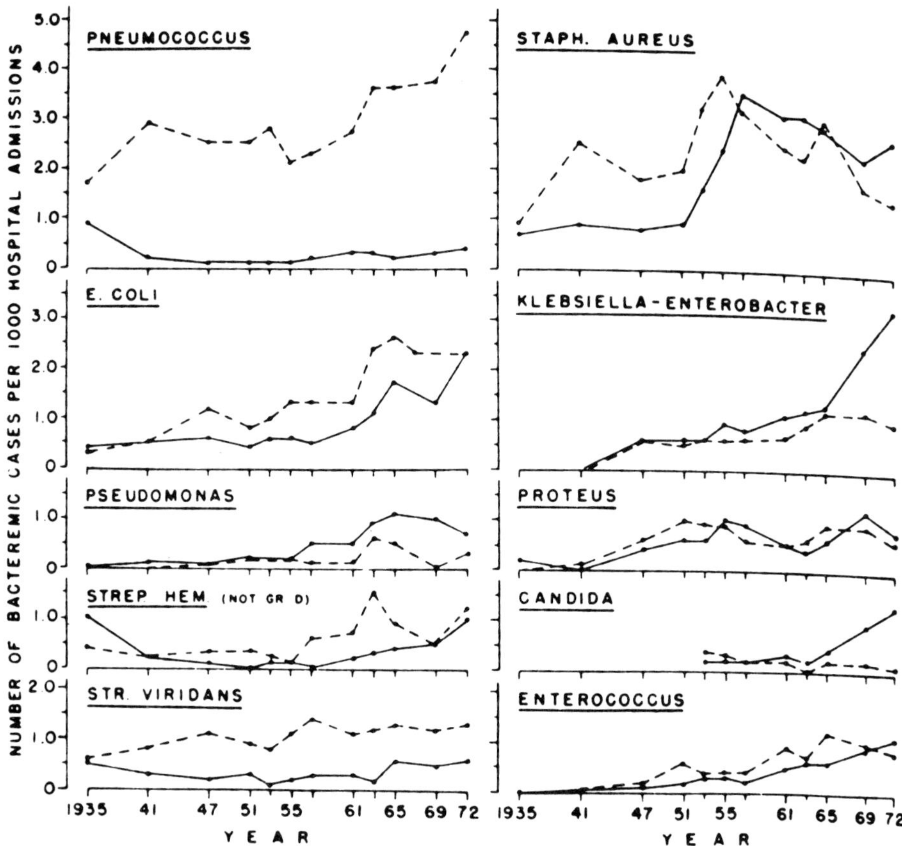

Fig. 7 *Incidence of nosocomial (—) and community-acquired (----) bacteremias at Boston City Hospital with various pathogens in each of 12 selected years between 1935 and 1972. Reproduced from McGowan et al (1975), with permission.*

As pointed out previously, on the average 15–45% of patients with gram-negative bacteremia do not survive hospitalization and the infection is a contributory or in many instances, the major cause of death (Fig. 1, Table 6). On the basis of the rates extracted from Dixon's data (1978), which are representative for most prospective studies (Table 6), it can be projected that each year in this country as many as 75 000 patients die with gram-negative bacteremia.

Case-control studies summarized by Maki (1981) have examined the economic impact of nosocomial bacteremias, most of which are caused by

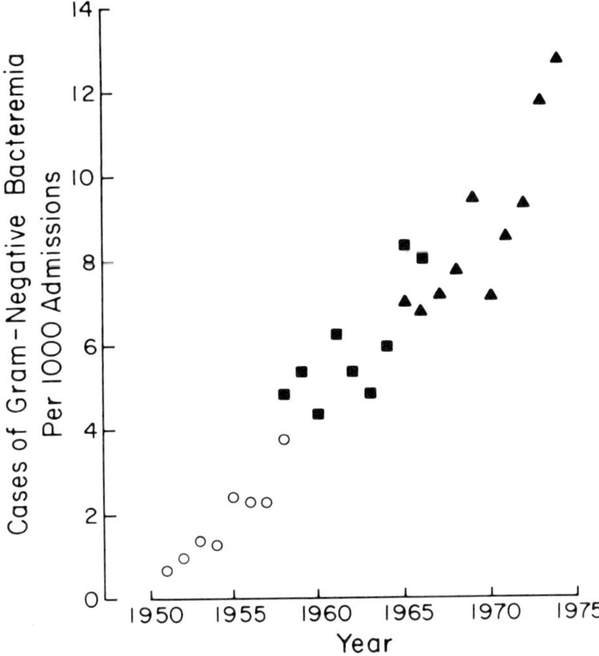

Fig. 8 *Secular trends in the incidence of gram-negative bacteremia in three American university hospitals: o = University of Illinois (McCabe and Jackson 1962a); ■ = University of Minnesota (Dupont and Spink 1969); ▲ = Boston University (Kreger et al 1980a).*

gram-negative bacilli. Independently derived data are in close agreement and suggest that in teaching hospitals, the development of a nosocomial bacteremia portends 14–19 days of additional hospitalization and commensurate extra hospital costs, compared with hospitalizations of matched control patients without bacteremia. Overall, these studies suggest that nosocomial gram-negative bacteremias account for up to one billion dollars extra health care costs each year in the United States.

5.2. Epidemic nosocomial bacteremias

As shown in Figure 4, the frequency of reported epidemics of nosocomial bacteremia has also increased greatly over the past 15 years, especially since 1970. The rise beginning in 1970 is very likely real and undoubtedly repre-

TABLE 6 *Incidence and mortality of gram-negative bacteremia, association with endotoxic shock, and shock-associated mortality**

Hospital, reported location	Years inclusive	Incidence (cases/1000 patients) in types of hospital			Proportion (%) with endotoxic shock	Case-fatality (%)	
		Municipal or federal	University	Community		Overall	Cases with shock
Boston City, Boston (McGowan et al 1975)	1972	11.5		N.S.	46	N.S.	
University of Illinois, Chicago (McCabe and Jackson, 1962a,b)	1951-1958		2.8		37	42	49
University of Minnesota, Minneapolis (Dupont and Spink 1969)	1958-1966		4.9		46	50	82
Stanford University, Palo Alto (Freid and Vosti 1968)	1959-1966		1.9		34	36	66
Peter Bent Brigham, Boston (Myerowitz et al 1971)	1966-1967		10.7		N.S.	26	N.S.
Boston University, Boston (Kreger et al 1980a, 1980b)	1965-1974		8.9		44	25	47
University of Colorado, Denver (Weinstein et al 1983a, 1983b)	1975-1977		6.4		42	48	N.S.
St. Mary's, Madison, WI (Scheckler 1977)	1970-1973			1.2	14	20	63
Hackensack, Hackensack, NJ (Setia and Gross 1977)	1974-1975			3.0	N.S.	27	N.S.
Four Charleston Hospitals, SC (Bryan et al 1983b)	1977-1981			3.9	16	36	64

* From published series providing incidence data and information on the frequency of associated endotoxic shock, shock-associated and overall mortality.

N.S. = not stated.

sents the aging population and the ever-increasing exposure of patients to immunosuppressive drugs, antibiotics, surgery, and invasive devices and procedures of all forms; it also reflects the trend of the past decade of treating virtually all critically ill patients in special units. It probably also indicates greatly awareness of nosocomial infection as a major health care problem, especially bacteremias related to infusion therapy (Maki 1981).

Case-fatality rates of nosocomial bacteremia in epidemics (mean for 86 outbreaks of bona fide bacteremia, 25%) have been considerably lower on the average than those reported from most centers for endemic bacteremias (Maki 1981). Except for outbreaks involving neonates that have been associated with exceptionally high mortality (mean for 24 outbreaks, 54%), this differential mortality obtained for most pathogens, including *P. aeruginosa*. Two important observations account for the disparate mortalities: (a) patients who become bacteremic in an epidemic tend to be younger and are less likely to have a fatal underlying disease than patients with endemic nosocomial bacteremia, and (b) in contrast to endemic bacteremias, most of which are secondary to a local infection (Fig. 6), the vast majority of epidemic cases are primary bacteremias; nearly two-thirds of the time the source is external to the patient (e.g., an intravenous infusion or a hemodialysis machine) and can therefore be immediately and totally excluded, if identified. The critical importance of this last observation is underscored by noting that even when bacteremia due to a vascular cannula occurs endemically, the case-fatality rate (19%) is much lower than when bacteremia originates from a natural local infection (mean for all other sites, 39.7%, $p < 0.001$; Maki 1981). In the 1970–71 nationwide epidemic due to contaminated infusion products of one U.S. manufacturer, the case-fatality of *Enterobacter* septicemia in 258 analyzed cases, 13.4% (Maki et al 1976), was considerably lower than the 35.8% mortality for endemic *Enterobacter* bacteremias in the Johns Hopkins Hospital between 1968 and 1974 (Spengler et al 1978).

The full economic impact of epidemic bacteremias is unknown. The direct consequences in terms of added health care costs are certainly considerably less than the projected costs of endemic bacteremias, but the aftermath of adverse publicity and litigation in epidemics may well be much greater. Despite their lamentable sequelae in terms of human suffering, epidemics have yielded important gains for preventing hospital-related bacteremia: our knowledge of the epidemiology of nosocomial infection, particularly bacteremias related to invasive devices, has been greatly expanded, and epidemics have enhanced general awareness that nosocomial infection, particularly bacteremia, is a very real risk of modern hospital care, but a risk that can be greatly reduced (Maki 1981).

6. NATURAL HISTORY OF GRAM-NEGATIVE SEPSIS AND RISK FACTORS FOR DEATH

6.1. Risk factors for a fatal outcome

On the basis of fragmentary data available from the pre-antibiotic era, it is probable that approximately 40% of patients with gram-negative septicemia died. It is often difficult in an individual case to determine the exact contribution of a bacteremic infection to a fatal outcome, especially when the patient already had a fatal underlying disease. However, case-control studies indicate that a hospitalized patient's risk of dying is increased 3.8 to 14 times when nosocomial bacteremia is present, compared with fatality rates of non-bacteremic control patients carefully matched for underlying diseases (Maki 1981), and it seems reasonable to surmise that gram-negative bacteremia, with or without endotoxic shock, is associated causally with substantial mortality.

Most of the published experiences with gram-negative bacteremia include analyses of factors predictive of a fatal outcome (Table 2) (Bryant et al 1971; Dupont and Spink 1969; Freid and Vosti 1968; Hodgin and Sanford 1965; Kreger et al 1980b; Lewis and Fekety 1969; Maiztegui et al 1965; McCabe and Jackson 1962b; McHenry et al 1975; Myerowitz et al 1971; Scheckler 1977; Setia and Gross 1977; Weil et al 1964; Weinstein et al 1983b). As pointed out previously, the McCabe and Jackson (1962b) system for classifying severity of underlying diseases has been found to be highly predictive of outcome in every subsequent study of gram-negative bacteremia. The most recent series – published since 1977 – show that bacteremic patients with ultimately fatal underlying diseases have 2–3-fold increased mortality and those with rapidly fatal diseases, 3–5-fold increased mortality, compared with case-fatality rates in bacteremic patients with nonfatal underlying diseases (Bryan et al 1983b; Kreger et al 1980b; Scheckler 1977). Approximately one-half of the deaths occur within the first 24 hours and 90% within 1 week of the onset of sepsis (Bryan et al 1983b; Kreger et al 1980a). Clearly, gram-negative bacilli, especially when they produce bacteremia associated with endotoxic shock, are prototypic opportunistic pathogens, and the pre-morbid physiologic status of the infected patient is a powerful predictor of clinical outcome.

Beyond severity of underlying disease, every one of the studies cited found survival in gram-negative bacteremia to also be very heavily influenced by whether or not endotoxic shock occurred (Fig. 1). Even in patients with nonfatal underlying diseases, shock greatly increases the case-fatality, as much as 4-fold (Kreger et al 1980b).

In terms of clinical risk factors that have been reported to be predictive of a fatal outcome (Table 2), most of these, including advanced age, fatal underlying disease, prior corticosteroid or antimicrobial therapy, sepsis deri-ving from the lung or gastrointestinal tract, *P. aeruginosa* or polymicrobial bacteremia, and inappropriate antimicrobial therapy, are also closely related to the development of endotoxic shock. Many of the factors associated with increased mortality from gram-negative bacteremia are strongly interrelated, and all of the cited studies reporting risk factor analyses (except for one) used univariate statistical methods. Multivariate techniques will also be re-quired to clarify the independence and relative importance of individual factors predictive of a fatal outcome. Although Weinstein et al (1983b) were the first to use multivariate techniques to identify predictors of mortality, they did not separate gram-negative bacteremias from gram-positive and fungal bloodstream infections in their analysis.

A hypothermic response to gram-negative sepsis has also been reported by many investigators as having very ominous portent; few patients who do not show a febrile response survive gram-negative sepsis (Bryant et al 1971; Gleckman and Hibert 1982; McCabe and Jackson 1962b; Weinstein et al 1983b). It might be argued whether either endotoxic shock or hypothermia should be regarded as true risk factors for fatal gram-negative sepsis in that, strictly speaking, prognostic factors should probably be pre-existent condi-tions, definable prior to the development of gram-negative sepsis. The vast majority of patients developing endotoxic shock or showing a hypothermic response with gram-negative sepsis are advanced in age or have pre-existent serious underlying conditions (e.g. fatal underlying diseases, renal or other major organ failure, previous corticosteroid therapy) that are powerful risk factors for development of endotoxic shock and a fatal outcome. Shock or hypothermia is simply an index of profoundly altered physiology that itself augers a very poor outcome with gram-negative sepsis.

6.2. Management of gram-negative sepsis

It is beyond the scope of this chapter to discuss in detail the clinical management of gram-negative sepsis, particularly the optimal selection and early use of systemic antimicrobial therapy, which should always be administered in maximal doses parenterally, and the vital measures necessary to support the failing circulation; both of these issues are addressed in detail in Chapters 5–8 and in Chapter 12, Volume 2. However, one facet of management bears emphasis.

In an individual case, *every effort must be made to identify the source of bacteremia and, if possible, to remove that source*, such as by surgically draining an abscess, decompressing an obstructed ureter or common bile duct, or closing an intestinal perforation; removing an infected intravascular catheter, possibly even resecting a suppurative vein segment (Maki 1982); discontinuing an intravenous or intraarterial infusion that may contain contaminants; or removing a heavily contaminated ventilator or hemodialysis machine. Such measures are crucial to survival and in many instances, such as sepsis from a contaminated infusion, far supersede antimicrobial therapy in therapeutic importance.

Harris and Cobbs (1973) described 20 patients with gram-negative bacteremia and sepsis persisting for 7 days or longer despite appropriate antimicrobial therapy; in every case a deep abscess or an endovascular focus of infection was shown to be the cause of refractory bacteremia. During a nationwide outbreak of gram-negative septicemia due to intrinsic contamination of intravenous solutions of one manufacturer in 1970–71, 17 of 19 patients in one hospital treated with gentamicin, to which the epidemic organisms were sensitive, remained clinically septic, continued to have positive blood cultures after 24 hours or more of gentamicin therapy, and did not show improvement until their infusions were fortuitously or intentionally removed (Maki et al 1976). Ledingham and McArdle (1978) have ascribed a two-fold improvement in survival of patients with septic shock from peritonitis to a policy of immediate exploratory laparotomy when an intraabdominal source of infection was suspected. Weinstein et al (1983b) found in a recent prospective study of 500 patients with bacteremia that failure to treat a removable focus of infection by surgery or other mechanical intervention increased the case-fatality rate nearly two-fold.

In most cases, the local gram-negative infection giving rise to bacteremia and a clinical picture of sepsis, with or without endotoxic shock, is patently apparent clinically, but if it is not, a picture of gram-negative sepsis occurring in a patient with one or more risk factors for gram-negative sepsis originating from a specific site (Table 6) should focus diagnostic efforts at identifying occult infection at that site. Conversely, recovery of certain species in blood cultures (Table 6) points strongly towards specific sites of local infection; for example, *B. fragilis* almost always denotes an intraabdominal source. *Enterobacter* species, particularly *E. agglomerans* or *E. cloacae, Flavobacterium* species, *P. cepacia* or *Citrobacter*, cultured from the blood of a patient receiving infusion therapy, should raise immediate suspicion of contaminated infusate and may well signal an epidemic; appropriate case-control studies and microbiologic sampling of in-use intravenous fluid and medications and possibly of unused samples should immediately be undertaken (Maki 1981).

As noted, approximately 40% of patients with gram-negative bacteremia in the pre-antibiotic era died. Review of the longitudinal studies of McCabe and his coworkers of gram-negative bacteremia in two hospitals over the past 30 years (Kreger et al 1980b; McCabe and Jackson 1962b) and secular trends in gram-negative bacteremia in various other U.S. centers (Table 6) indicate that there has been significant improvement in the outcome of treated bacteremic gram-negative infections over the past 30 years. It is probable that the enormous advances in antimicrobial therapy, resulting in increasingly potent antimicrobial drugs effective against virtually all gram-negative bacilli, including *P. aeruginosa*, account for much of this trend. If the diagnosis of gram-negative sepsis is made promptly and optimal therapy is instituted, including *early* administration of effective anti-infective drugs (Bryan et al 1983b), efforts to identify and extirpate removable foci of infection, and supportive measures for the circulation and failing organs, at least three-fourths of all patients with gram-negative sepsis, irrespective of the underlying disease, should survive; in patients with endotoxic shock, more than half should survive with optimal therapy.

Because of marked differences in populations treated, patients in community hospitals in the United States today are clearly much less likely to develop gram-negative sepsis with endotoxic shock (14–16%, Table 6) than patients in large university referral centers (42–44%). However, patients developing endotoxic shock in university hospitals, which have far greater experience in dealing with this entity, appear to be considerably more likely to survive their bacteremic infection (53% at Boston University Hospital) than the relatively few patients developing endotoxic shock in smaller community hospitals (36–37% in four reporting community hospitals, Table 6).

6.3. The future

We believe that further improvement in the survival rates of patients with gram-negative sepsis and particularly, endotoxic shock, will require major progress in our ability to prevent or treat the subcellular aberrations which predictably occur in endotoxic shock and result in multiple organ failure and death (Mizock 1984); to achieve this, new biochemical and immunologic interventions will have to be used. As noted, Ziegler et al (1982) demonstrated a two-fold improvement of survival rates of patients with gram-negative endotoxic shock treated with the experimental J5 antiserum, containing high titer antibodies against the core gram-negative cell wall lipopolysaccharide antigen, over patients receiving appropriate antimicrobial therapy and optimal treatment in all other respects but not the hyperimmune serum. Better understanding of the biochemical causes of multiple organ failure occurring in endotoxic shock should permit additional innovative therapeutic approaches to prevent organ failure and even further improve our ability to treat gram-negative endotoxic shock.

7. PREVENTION OF GRAM-NEGATIVE BACTEREMIA

Certainly not all gram-negative bacteremias can be prevented with the current state of knowledge (McGowan et al 1977; Spengler and Greenough 1978), especially those occurring in immunologically severely compromised patients. However, many gram-negative bacteremias occur in hospitalized patients who are not intrinsically immunologically compromised. This point is underscored on considering the extraordinarily high rates of nosocomial bacteremia reported to occur in patients in critical care units, ranging up to 350 cases per 1000 patients (Daschner et al 1982; Goldmann et al 1981; Maki 1981; Miller et al 1973; Northey et al 1974). As a rule, these patients are not immunologically compromised, but they are invariably subjected to multiple invasive devices and procedures and often, surgery.

Measures for prevention of infection in general, including gram-negative bacteremia, encompass: (a) striving to block or at least delay colonization of vulnerable areas by gram-negative pathogens; (b) improving the safety of invasive procedures and devices which greatly amplify the risk of infection; and (c) augmenting immune defenses by the use of vaccines or possibly by providing passive immunity (Maki 1981; Stamm et al 1977).

108

Greater attention to high priority aseptic practices could materially reduce the incidence of local nosocomial gram-negative infections and in the process, secondary bacteremia (Goldmann et al 1981; Iffy et al 1979). More rapid diagnosis and prompter and more effective treatment of local gram-negative infections could also prevent many secondary bacteremias that might otherwise develop. However, the potential for prevention is probably greatest for epidemic nosocomial bacteremias, the majority of which are related to exposure to invasive devices, to heavy levels of environmental contaminants – often from a common source – or both (Maki 1981).

Infection control guidelines have been published dealing with most of the high-risk areas, devices or procedures in the hospital where lapses in aseptic technique or contamination can produce gram-negative bacteremia: isolation procedures, handwashing, cutaneous antisepsis, disinfection and sterilization, urethral catheters, infusion therapy, hyperalimentation, hemodynamic arterial pressure monitoring systems, respiratory therapy, surgery, the newborn nursery, hemodialysis, environmental microbiologic surveillance, and prophylaxis for endocarditis with invasive procedures.* Moreover, the CDC has recently published detailed guidelines for prevention of nosocomial infection in each area developed by panels of national experts.**

Continuous surveillance of all nosocomial infections as an infection control measure remains a controversial issue in the present national climate of intense commitment to cost containment in health care. However, bacteremia connotes unimpeachable laboratory confirmation of infection, and surveillance data on bacteremic infections are as a rule highly reliable epidemiologically and provide an accurate measure of the hospital's problem with nosocomial infection in general. Whereas the value of continuous surveillance of all nosocomial infections as a control measure remains unclear, surveillance of bacteremias, much simpler and far less expensive, may be more worthwhile. However, to achieve optimal results, as noted earlier in this chapter, it is imperative that all blood isolates be routinely identified as to species, that complete antimicrobial susceptibility testing with standardized methods be performed routinely on all blood isolates, and that an isolate from each clinically relevant bacteremia be routinely saved by the laboratory for at least 12–24 months for future study should an epidemic occur.

---

* For individual citations to published control measures, see Maki (1981).

** Published in the journal *Infection Control*, between 1981 and 1985; some have been cited earlier in this review.

Maki (1981) has suggested that advances in the following areas could substantially reduce the overall incidence of endemic nosocomial bacteremia, which would implicitly afford major reductions in gram-negative bacteremic infections as well:

a. Techniques to allow more rapid and precise diagnosis of local infections, permitting quicker and more effective treatment to prevent secondary bacteremias that might otherwise develop.

b. Studies to better define the epidemiology of endemic nosocomial infections, especially those caused by gram-negative bacilli, to develop more effective guidelines for prevention.

c. Major resources have been committed to study the efficacy of complex protective isolation systems for patients with acute leukemia, aplastic anemia and those who have undergone bone marrow transplantation (Buckner et al 1978). However, except for limited studies in burn patients (Burke et al 1977; Jarrett et al 1978), few studies have examined the efficacy of these measures to prevent or at least delay nosocomial colonization and thus reduce invasive gram-negative infections in neonates, trauma patients, and others confined to critical care units. Studies are now needed to evaluate the efficacy of various forms of protective isolation of these patients who usually have a limited period of vulnerability.

d. Improvements in the design of devices and development of new techniques to enhance their safety.

e. More judicious use of broad-spectrum systemic antibiotics, especially in high-risk patients who are immunocompromised or are heavily exposed to invasive devices in the setting of a critical care unit. Antibiotics are vastly overused for surgical prophylaxis in most hospitals, especially in the postoperative period (Kunin 1985).

f. Many physicians are unaware of basic precepts of infection control and in most hospitals nurses are far better informed and are the most effective force for infection control. Innovative programs are needed for physicians and other health care professionals to enhance awareness of nosocomial infection and to better communicate knowledge of infection control, especially with respect to the use of devices.

g. Vaccines directed against the shared core cell wall antigens of the enteric gram-negative bacilli (McCabe et al 1973) or against the prevalent serogroups of *P. aeruginosa* (Alexander et al 1971; Young et al 1973) already show promise and must continue to be vigorously pursued.

8. SUMMARY

Gram-negative sepsis associated with endotoxic shock is one of the most serious acute infections yet producing major morbidity and mortality in the developed countries. It is clear from longitudinal studies in a number of United States centers that the incidence of gram-negative bacteremia in American hospitals has risen progressively over the past four decades and that currently approximately 5 patients per 1000 U.S. hospital admissions either have gram-negative bacteremia on admission or, more commonly, develop nosocomial gram-negative bacteremia during hospitalization. Overall case-fatality rates for gram-negative bacteremia are in the range of 25–40%; endotoxic shock, present in approximately one-third of cases, increases the mortality to 45%–65%. It can be estimated that each year in the United States there are at least 200 000 cases of gram-negative bacteremia, resulting in approximately 75 000 deaths.

Hypotension alone is not synonymous with endotoxic shock. We suggest the following criteria for making a diagnosis of endotoxic shock: (a) documented gram-negative septicemia or, if bacteremia is not confirmed, a symptomatic major local gram-negative infection – that is, clear-cut gram-negative sepsis; and (b) evidence of generalized hypoperfusion, best characterized by lactemia and metabolic acidosis, and clinical signs of organ failure from inadequate perfusion, such as altered mental status, oliguria, a high (or low) cardiac output associated with an incommensurately low systemic vascular resistance, noncardiogenic pulmonary edema, ileus, or coagulopathies, particularly disseminated intravascular coagulation.

In most cases of gram-negative endotoxic shock, the major exceptions being cases originating from intravascular devices or deep intraabdominal abscesses, the signs and symptoms of the local infection giving rise to the picture of sepsis predominate. Fever, associated with the chills or shaking rigors, and hyperventilation are almost universal. Jaundice and nonspecific gastrointestinal or neurologic symptoms also commonly occur and may divert attention away from diagnostic consideration of sepsis.

Leukocytosis and morphologic abnormalities of granulocytes in the peripheral blood smear, particularly a marked 'shift to the left', vacuolization of neutrophils, and intraneutrophilic Döhle bodies are extremely common, as are coagulation abnormalities, often indicative of disseminated intravascular coagulation. Hypoxemia and acid–base disorders are also very common, as is hyperbilirubinemia. Oliguria and azotemia usually accompany shock

111

and have ominous prognostic significance.

Most importantly, for an early diagnosis of gram-negative sepsis a high degree of suspicion is warranted, especially in patients with altered host defenses or in a vulnerable setting, such as patients with one or more invasive devices or those requiring critical care unit support.

Laboratory confirmation of gram-negative sepsis is predicated upon identifying the local source of an infection, both clinically and microbiologically. If bacteremia is present, which it usually is, the infecting organisms should be recovered in blood cultures. Gram-stained smears of urine, tracheal secretions, or of exudate from the area of purulence will show the infecting organism in large numbers in over 95% of cases originating from the urinary tract, lower respiratory tract, skin or soft tissues, or an abscess. Culture of at least 30 ml of blood from adults (two 15 ml specimens or three 10 ml specimens) will yield the bloodstream pathogen, if present, over 99% of the time. Recovery of enteric gram-negative bacilli, *P. aeruginosa*, or anaerobic gram-negative bacilli nearly always reflects true bacteremia, rather than contaminated cultures.

The majority of gram-negative bacteremias originating in the community develop secondarily, as a complication of an identifiable local infection, most commonly of the urinary tract, an intraabdominal focus, or of the skin and soft tissues. In contrast, although many gram-negative bacteremias originating in the hospital also derive from these sites, nearly 25% of endemic gram-negative bacteremias are primary bacteremias; most primary nosocomial cases originate from intravascular devices of various forms.

E. coli remains preeminent, both in community-acquired and nosocomial cases of gram-negative endotoxic shock, but accounts for a considerably higher proportion of cases originating in the community. Bacteremias caused by *H. influenzae* type b, *N. meningitidis*, *N. gonorrhoeae*, and *Salmonella* are seen almost exclusively in community-acquired cases. In contrast, bacteremias caused by *Klebsiella*, *P. aeruginosa*, other *Pseudomonas* species, *Serratia*, *Acinetobacter*, *Flavobacterium*, and other 'water organisms' are relatively infrequent in community-acquired cases, but account for over one-half of all nosocomial bloodstream infections. Bacteremia is more likely to occur with local infections caused by *Bacteroides* species, *Serratia*, *Acinetobacter*, and *Providencia*.

In the past 20 years there have been numerous outbreaks of nosocomial bacteremia in hospitals; between 1965 and 1978, 97 epidemics were reported in the world literature. Over three-fourths of these outbreaks were caused by aerobic gram-negative bacilli, particularly *Klebsiella* and *Enterobacter* species, *S. marcescens*, *P. aeruginosa*, and especially, other pseudomonads such as *P. cepacia*. Nearly 80% of all outbreaks were primary bacteremias, deriving from infusion therapy in some form, hemodialysis, or other invasive devices involving the bloodstream. There have been a dozen outbreaks of acute pyrogenic reactions in which bacteremia could not be documented; investigations have shown, however, that in these outbreaks, which were associated with hemodialysis, cardiac catheterization, or administration of a common blood product, nonviable pyrogens, shown to represent endotoxin, were introduced into the patients' bloodstreams.

Since 1970 there have been six outbreaks of national scope involving multiple hospitals and deriving from manufacturer-related contamination of a nationally distributed product. Intrahospital surveillance networks, such as the National Nosocomial Infection Surveillance study of the CDC, may well be the most effective means of detecting these common-source outbreaks early.

Major risk factors for development of gram-negative sepsis and in most cases, endotoxic shock as well, include advanced age, a fatal underlying disease, prior corticosteroid or antimicrobial therapy, *P. aeruginosa* bacteremia, bacteremia originating from peritonitis, pyohydronephrosis, obstructive cholangitis, or gram-negative pneumonia, or treatment with antibiotics ineffective against the infecting organism in vitro ('inappropriate antimicrobial therapy').

Patients developing endotoxic shock have a three- to four-fold increased case-fatality. Other patients at high risk of dying with gram-negative sepsis include patients with a fatal underlying disease; showing a hypothermic response to the infection; with granulocytopenia or multiple organ failure (especially renal or respiratory failure); who have received prior antimicrobial or corticosteroid therapy; with infection originating in the hospital; with an unknown, respiratory tract, or intraabdominal source of infection; with *P. aeruginosa* or polymicrobial bacteremia; or who are given inappropriate antimicrobial therapy.

In an individual case of probable gram-negative sepsis, every effort must be made to identify the source of bacteremia, to permit the most specific antimicrobial therapy and if possible, to extirpate that source, such as by surgical drainage of an abscess, decompression of an obstructed ureter, bile duct, or closure of an intestinal perforation, or by removing an infected intravascular catheter or contaminated infusion. Failure to treat a removable focus of infection by surgery or other mechanical intervention increases the case-fatality rate two-fold.

Further improvement in the survival of patients with gram-negative sepsis and particularly endotoxic shock will require major advances in our ability to prevent or treat the subcellular aberrations which predictably occur in endotoxic shock and result in multiple organ failure and death. The demonstrated successful use in human gram-negative endotoxic shock of immune serum, with a high titer of antibodies against the core gram-negative lipopolysaccharide antigens, is a major step forward in this direction.

Clearly, not all cases of gram-negative sepsis, especially those associated with endotoxic shock, can be prevented with the current state of knowledge. However, many cases occur in hospitalized patients who are not intrinsically immunocompromised; these are causally related to invasive devices or surgery. More sensitive and reliable techniques for rapid diagnosis, leading to earlier and more specific treatment of local gram-negative infections, could prevent many secondary bacteremias. Studies are needed to better define the epidemiology of endemic nosocomial gram-negative infections, to guide development of more effective measures for prevention of infections related to devices and surgery. Improvements in the design of devices and development of new techniques to enhance their safety are also important, while studies to evaluate the efficacy of protective isolation in neonates, trauma patients, and others confined to critical care units – who have limited periods of vulnerability – also seem of high priority. Broad-spectrum systemic antibiotics that promote colonization and superinfection by gram-negative bacilli, especially in immunocompromised patients and in patients undergoing surgery or confined to critical care units, must begin to be used more judiciously. Innovative programs are urgently needed to communicate more effectively existent knowledge of infection control – which is not being consistently applied – to physicians and other health care professionals, especially with respect to the use of devices.

REFERENCES

Aber RC, Garner JS (1981) Postoperative wound infections. In: Wenzel RP (Ed) *Handbook of Hospital Acquired Infection*, pp 303-316. CRC Press, Boca Raton, FL.

Aber RC, Mackel DC (1981) Epidemiologic typing of nosocomial microorganisms. *Am. J. Med. 70*, 899-905.

Alexander JW, Fisher MW, McMillan BG (1971) Immunological control of Pseudomonas infection in burn patients: a clinical evaluation. *Arch. Surg. 102*, 31-39.

Anagnostakis D, Kamba A, Petrochilou V, Arseni A, Marsaniotis N (1975) Risk of infection associated with umbilical vein catheterization: a prospective study in 75 newborn infants. *J. Pediatr. 86*, 759-765.

Anderson ET, Young LS, Hewitt WL (1976) Simultaneous antibiotic levels in 'break through' gram-negative rod bacteremia. *Am. J. Med. 61*, 493-497.

Andrew DJ, Johnson SE (1970) Acute suppurative cholangitis, a medical and surgical emergency: a review of 10 years experience emphasizing early recognition. *Am. J. Gastroenterol. 54*, 141-154.

Anhalt JP (1978) New or experimental approaches to detection of bacteremia. In: Washington JA III (Ed) *The Detection of Septicemia*, pp 109-144. CRC Press, West Palm Beach.

Baine WB, Gangarosa EJ, Bennett JV, Barker WH Jr (1973) Institutional salmonellosis. *J. Infect. Dis. 128*, 357-360.

Band JD, Maki DG (1979) Safety of changing intravenous delivery systems at longer than 24-hour intervals. *Ann. Intern. Med. 91*, 173-178.

Banks JG, Foulis AK, Ledingham IMcA, MacSween RNM (1982) Liver function in septic shock. *J. Clin. Pathol. 35*, 1249-1252.

Bauer E, Densen P, Faxon D, Kloster C, Melidossian C, Kundsin R, Ryan P, Fiumara N (1979) Endotoxic reactions associated with the reuse of cardiac catheters – Massachusetts. *Morbid. Mortal. Weekly Rep. 28*, 25-27.

Baxter CR (1972) Surgical management of soft tissue infections. *Surg. Clin. North Am. 52*, 1483-1499.

Beller FK, Douglas GW (1973) Thrombocytopenia indicating gram-negative infection and endotoxemia. *Obstet. Gynecol. 41*, 521-524.

Bodey GP (1970) Epidemiological studies of Pseudomonas species in patients with leukemia. *Am. J. Med. Sci. 260*, 82-89.

Bryan CS, Reynolds KL (1984a) Community-acquired bacteremic urinary tract infection: epidemiology and outcome. *J. Urol. 132*, 490-493.

Bryan CS, Reynolds KL (1984b) Hospital-acquired bacteremic urinary tract infection: epidemiology and outcome. *J. Urol. 132*, 494-498.

Bryan CS, Reynolds KL (1984c) Bacteremic nosocomial pneumonia. Analysis of 172 episodes from a single metropolitan area. *Am. Rev. Resp. Dis. 129*, 668-671.

Bryan CS, Dew CE, Reynolds KL (1983a) Bacteremia associated with decubitus ulcers. *Arch. Intern. Med. 143*, 2093-2095.

Bryan CS, Reynolds KL, Brenner ER (1983b) Analysis of 1,186 episodes of gram-negative bacteremia in non-university hospitals: the effects of antimicrobial therapy. *Rev. Infect. Dis. 5*, 629-638.

Bryan CS, Reynolds KL, Moore EE (1984c) Bacteremia in obstetrics and gynecology.

Obstet. Gynecol. 64, 155-158.

Bryant RE, Hood AF, Hood CE, Koenig MG (1971) Factors affecting mortality of gram-negative rod bacteremia. *Arch. Intern. Med. 127*, 120-128.

Buckner CD, Clift RA, Sanders JE, Meyers JD, Counts GW, Farewell VT, Thomas ED and the Seattle Marrow Transplant Team (1978) Protective environment for marrow transplant recipients: a prospective study. *Ann. Intern. Med. 89*, 893-901.

Büller HR, Bolweck C, Ten Cate J, Roos J, Kahle LH, Ten Cate JW (1982) Postoperative hemostatic profile in relation to gram-negative septicemia. *Crit. Care Med. 10*, 311-315.

Burbrick MP, Hitchcock CR (1979) Necrotizing anorectal and perineal infections. *Surgery 86*, 655-662.

Burke JF, Quinby WC, Bondoc CC, Sheehy EM, Moreno HC (1977) The contribution of a bacterially isolated environment to the prevention of infection in seriously burned patients. *Ann. Surg. 186*, 377-387.

Carey RB (1984) Clinical comparison of the Isolator 1.5 microbial tube and the BACTEC radiometric system for detection of bacteremia in children. *J. Clin. Microbiol. 19*, 634-638.

Casali RE, Tucker WE, Petrino RA, Westbrook KC, Read RC (1980) Postoperative necrotizing fasciitis of the abdominal wall. *Am. J. Surg. 140*, 787-790.

Cavanagh D, Knuppel RA, Shepherd JH, Anderson R, Rao PS (1982) Septic shock and the obstetrician/gynecologist. *South. Med. J. 75*, 809-813.

Centers for Disease Control (1970) Epidemiologic aspects of nursery salmonellosis. *National Nosocomial Infections Study Report May 1970*, 15-16.

Centers for Disease Control (1974) Nosocomial salmonellosis – 2 outbreaks, *National Nosocomial Infections Study Quarterly Report, First and Second Quarters, 1973*, 13-14.

Centers for Disease Control (1981) Guidelines for prevention of catheter-associated urinary tract infections. *Infect. Control 2*, 1-5.

Centers for Disease Control (1982a) Guidelines for prevention of nosocomial pneumonia. *Infect. Control 3*, 327-333.

Centers for Disease Control (1982b) Guidelines for prevention of surgical wound infections. *Infect. Control 3*, 188-196.

Cohen P, Gardner FH (1966) Thrombocytopenia as a laboratory sign and complication of gram-negative bacteremic infection. *Arch. Intern. Med. 117*, 113-124.

Conn HO, Fessel JM (1971) Spontaneous bacterial peritonitis in cirrhosis: variations on a theme. *Medicine (Baltimore) 50*, 161-197.

Corrigan JJ Jr (1977) Medical progress: heparin therapy in bacterial septicemia. *J. Pediatr. 91*, 695-700.

Corrigan JJ Jr, Ray WL, May N (1968) Changes in blood coagulation system associated with septicemia. *N. Engl. J. Med. 279*, 851-856.

Cowett RM, Peter G, Hakanson DO, Oh W (1976) Reliability of bacterial culture of blood obtained from an umbilical artery catheter. *J. Pediatr. 88*, 1035-1036.

Cruse PJE, Foord R (1980) The epidemiology of wound infection. A ten-year prospective study of 62,939 wounds. *Surg. Clin. North Am. 60*, 27-40.

Daschner FD, Frey P, Wolff G, Baumann PC, Suter P (1982) Nosocomial infections in intensive care wards. A multicenter prospective study. *Intensive Care Med. 8*, 5-9.

Dietzman DE, Fischer GW, Schoenknecht FD (1974) Neonatal *Escherichia coli* septicemia – bacterial counts in blood. *J. Pediatr. 85*, 128 130.

Dixon RE (1978) Effects of infections on hospital care. *Ann. Intern. Med. 89*, 749-753.

Dobkin JF, Miller MH, Steigbiegel N (1978) Septicemia in patients on chronic hemodialysis. *Ann. Intern. Med. 88*, 28-33.

Dorff GJ, Geimer NF, Rosenthal DR, Rytel MW (1971) Pseudomonas septicemia. *Arch. Intern. Med. 128*, 591-595.

Dupont HL, Spink WW (1969) Infections due to gram-negative organisms: an analysis of 860 patients with bacteremia at the University of Minnesota medical center. *Medicine (Baltimore) 48*, 307-332.

Edwards LD (1976) The epidemiology of 2056 remote site infections and 1,966 surgical wound infections occurring in 1,865 patients: a four-year study of 40,923 operations at Rush-Presbyterian-St. Luke's Hospital, Chicago. *Ann. Surg. 184*, 758-766.

Emerson WA, Zieve PD, Krevans JR (1970) Hematologic changes in septicemia. *Johns Hopkins Med. J. 126*, 69-76.

Favero MS, Petersen NJ (1977) Microbiologic guidelines for hemodialysis systems. *Dialysis Transplant. 6*, 34-36.

Felices FJ, Hernandez JL, Ruiz J, Mesguer J, Gomes JA, Molina E (1979) Use of the central venous pressure catheter to obtain blood cultures. *Crit. Care Med. 7*, 78-82.

Felty AR, Keefer CS (1924) *Bacillus coli* sepsis: a clinical study of twenty-eight cases of blood stream infection by the colon bacillus. *J. Am. Med. Assoc. 82*, 1430-1433.

Finland M, Jones WF Jr, Barnes MW (1959) Occurrence of serious bacterial infections since introduction of antibacterial agents. *J. Am. Med. Assoc. 170*, 2188-2197.

Fowler AA, Hamman RF, Good JT, Benson KN, Baird M, Eberle DJ, Petty TL, Hyers TH (1983) Adult respiratory distress syndrome: risk with common predispositions. *Ann. Intern. Med. 98*, 593-597.

Franciosi RA, Favero BE (1972) A single blood culture for confirmation of the diagnosis of neonatal septicemia. *Am. J. Clin. Pathol. 57*, 215-219.

Freid MA, Vosti KL (1968) The importance of underlying disease in patients with gram-negative bacteremia. *Arch. Intern. Med. 121*, 418-423.

Galpin JE, Chow AW, Bayer AS, Guze LB (1976) Sepsis associated with decubitus ulcers. *Am. J. Med. 61*, 346-350.

Garibaldi RA (1981) Hospital acquired urinary tract infection. In: Wenzel RP (Ed), *Handbook of Hospital Acquired Infection*, pp 513-537. CRC Press, Boca Raton, FL.

Garibaldi RA, Britt MR, Coleman ML, Reading JC, Pace NL (1981) Risk factors for postoperative pneumonia. *Am. J. Med. 70*, 677-680.

Gleckman R, Hibert D (1982) Afebrile bacteremia. A phenomenon in geriatric patients. *J. Am. Med. Assoc. 248*, 1478–1481.

Gleckman R, Blagg N, Hibert D, Hall A, Crowley M, Pritchard A, Warren W (1982) Community-acquired bacteremic urosepsis in the elderly patients: a prospective study of 34 consecutive episodes. *J. Urol. 128*, 79–81.

Goldmann DA, Fulkerson CC, Dixon RE, Maki DG, Bennett JV (1978) Nationwide epidemic of septicemia caused by contaminated intravenous products. II. Assessment of the problem by a national nosocomial infection surveillance system. *Am. J. Epidemiol. 108*, 207-213.

Goldmann DA, Durbin WA Jr, Freeman J (1981) Nosocomial infections in a neonatal intensive care unit. *J. Infect. Dis. 144*, 449-460.

Goodpasture HC, Romig DA, Voth DW, Liu C, Brackett CE (1977) A prospective study of tracheobronchial bacterial flora in acutely brain-injured patients with and without antibiotic prophylaxis. *J. Neurosurg. 47*, 228-235.

Gorbach SL, Bartlett JG (1974) Anaerobic infections. *N. Engl. J. Med. 290*, 1177-1184.

Haley RW, Hooton TM, Culver DH, Stanley RC, Emori TG, Hardison CD, Quade D, Schachtman RH, Schaberg DR, Shah BV, Schatz GD (1981) Nosocomial infections in U.S. hospitals, 1975-1976: estimated frequencies by selected characteristics of patients. *Am. J. Med. 70*, 947-959.

Hall MM, Ilstrup DM, Washington JA II (1976) Effect of volume of blood cultured on detection of bacteremia. *J. Clin. Microbiol. 3*, 643-645.

Han T, Soral JE, Neter E (1967) Salmonellosis in disseminated malignant diseases: a seven-year review (1959-1965). *N. Engl. J. Med. 276*, 1045-1052.

Hansen SL, Hetemanski J, Stewart BJ (1983) Value of resin devices for the recovery of organisms from blood and other body fluids. *Abstr. Annu. Meet. Am. Soc. Microbiol. C63*, 322.

Harris JA, Cobbs CG (1973) Persistent gram-negative bacteremia: observations in twenty patients. *Am. J. Surg. 125*, 705-717.

Harris H, Wirtschafter D, Cassady G (1976) Endotracheal intubation and its relationship to bacterial colonization and systemic infection of newborn infants. *Pediatrics 56*, 816–820.

Hemming VG, Overall JC Jr, Britt MR (1976) Nosocomial infections in a newborn intensive-care unit: results of forty-one months of surveillance. *N. Engl. J. Med.*

294, 1310-1316.

Hermans PE, Washington JA II (1970) Polymicrobial bacteremia. *Ann. Intern. Med. 73*, 387-392.

Hodgin UG, Sanford JP (1965) Gram-negative rod bacteremia: an analysis of 100 patients. *Am. J. Med. 39*, 952-960.

Hooten TM, Haley RW, Culver DH (1980) A method for classifying patients according to their nosocomial infection risks associated with diagnoses and surgical procedures. *Am. J. Epidemiol. 111*, 556-573.

Hooten TM, Haley RW, Culver DH, White JW, Morgan WM, Carroll RJ (1981) The joint associations of multiple risk factors with the occurrence of nosocomial infection. *Am. J. Med. 70*, 960-970.

Ing AFM, McLean APH, Meakins JL (1981) Multiple-organism bacteremia in the surgical intensive care unit: a sign of intraperitoneal sepsis. *Surgery 90*, 779-786.

Iffy L, Kaminetzky HA, Maidman JE, Lindsey J, Arrata WSM (1979) Control of perinatal infection by traditional preventive measures. *Obstet. Gynecol. 54*, 403-411.

Jacob L (1909) Über Allgemeininfektionen durch *Bacterium coli* commune. *Deutsche Arch. Klin. Med. 97*, 303-347.

Jarrett F, Balish E, Moylan JA, Ellerbe S (1978) Clinical experience with prophylactic antibiotic bowel suppression in burn patients. *Surgery 83*, 523-527.

Jarvis WR, White JW, Munn VP, Mosser JL, Emori TG, Culver DH, Thornsberry C, Hughes J (1984) Nosocomial infection surveillance, 1983. *Morbid. Mortal. Weekly Rep. 33*, 955-2155.

Johanson WG, Pierce AK, Sanford JP (1969) Changing pharyngeal bacterial flora of hospitalized patients: emergence of gram-negative bacilli. *N. Engl. J. Med. 281*, 1137-1140.

Johanson WG Jr, Pierce AK, Sanford JP, Thomas GD (1972) Nosocomial respiratory infections with gram-negative bacilli: the significance of colonization of the respiratory tract. *Ann. Intern. Med. 77*, 701-706.

Kaplan RL, Sahn SA, Petty TL (1979) Incidence and outcome of the respiratory distress syndrome in gram-negative sepsis. *Arch. Intern. Med. 139*, 867-869.

Kaye D (1980) Urinary tract infection in the elderly. *Bull. N.Y. Acad. Med. 56*, 209-220.

Keane WF, Shapiro FL, Raiji L (1977) Incidence and type of infections occurring in 445 chronic hemodialysis patients. *Trans. Am. Soc. Artif. Intern. Organs 24*, 41-50.

Kiehn TE, Wong B, Edwards FF, Armstrong D (1983) Comparative recovery of bacteria and yeasts from lysis-centrifugation and a conventional blood culture system. *J. Clin. Microbiol. 18*, 300-304.

Kluge RM, Dupont HL (1973) Factors affecting mortality of patients with bacteremia. *Surg. Gynecol. Obstet. 137*, 267-269.

Koliner C, Boulanger M, McLean APH, Meakins JL (1980) Nonbacteremic clinical sepsis: an entity? *Crit. Care Med. 8*, 230.

Kolmos HJ, Andersen KEH (1979) Peritonitis with *Pseudomonas aeruginosa* in hospital patients treated with peritoneal dialysis. *Scand. J. Infect. Dis. 11*, 207-210.

Kraus ES, Spector DA (1983) Characteristics and sequelae of peritonitis in diabetics and non-diabetics receiving chronic intermittent peritoneal dialysis. *Medicine (Baltimore) 62*, 52-57.

Kreger BE, Craven DE, Carling PC, McCabe WR (1980a) Gram-negative bacteremia. III. Reassessment of etiology, epidemiology and ecology in 612 patients. *Am. J. Med. 68*, 332-343.

Kreger BE, Craven DE, McCabe WR (1980b) Gram-negative bacteremia. IV. Reevaluation of clinical features and treatment in 612 patients. *Am. J. Med. 68*, 344-355.

Krieger JN, Kaiser DL, Wenzel RP (1983) Urinary tract etiology of bloodstream infections in hospitalized patients. *J. Infect. Dis. 148*, 57-62.

Kunin CM (1985) The responsibility of the infectious disease community for the optimal use of antimicrobial agents. *J. Infect. Dis. 151*, 388-389.

Lang DJ, Kunz LJ, Martin AR, Schroeder SA, Thomson LA (1967) Carmine as a source of nosocomial Salmonellosis. *N. Engl. J. Med. 267*, 829-832.

Ledger WJ (1984) Infections of the female pelvis. In: Mandell GL, Douglas RG, Bennett JE (Eds), *Principles and Practice of Infectious Diseases*, pp 738-745. John Wiley and Sons, New York.

Ledger WJ, Norman M, Gee C, Lewis W (1975) Bacteremia on an obstetric-gynecologic service. *Am. J. Obstet. Gynecol. 121*, 205-212.

Ledingham IMcA, McArdle CS (1978) Prospective study of the treatment of septic shock. *Lancet 1*, 1194-1197.

Lee S, Schoen I, Malin A (1967) Comparison of use of alcohol with that of iodine for skin antisepsis in obtaining blood cultures. *Am. J. Clin. Pathol. 47*, 646-648.

LeFrock JL, Ellis CA, Weinstein L (1979a) The impact of hospitalization on the aerobic fecal microflora. *Am. J. Med. Sci. 277*, 269-274.

LeFrock JL, Ellis CA, Weinstein L (1979b) The relation between aerobic fecal and oropharyngeal microflora in hospitalized patients. *Am. J. Med. Sci. 277*, 275-280.

Lewis J, Fekety FR (1969) Gram-negative bacteremia. *Johns Hopkins Med. J. 19*, 106-111.

Lindberg RB, Moncrief JA, Mason AD Jr (1968) Control of experimental and clinical burn wound sepsis by topical application of sulfamylon compounds. *Ann. N.Y. Acad. Sci. 150*, 950-960.

Lorber B, Swenson RM (1974) Bacteriology of aspiration pneumonia: a prospective study of community and hospital-acquired cases. *Ann. Intern. Med. 81*, 329-331.

Lufkin EG, Silverman M, Callaway JJ, Glenchur H (1966) Mixed septicemias and gastrointestinal disease. *Am. J. Dig. Dis. 11*, 930-937.

MacFarlane JT, Finch RG, Ward MJ, Macrae AD (1982) Hospital study of adult community-acquired pneumonia. *Lancet 2*, 255-258.

MacGregor RR, Beaty HN (1972) Evaluation of positive blood cultures: guidelines for early differentiation of contaminated from valid positive cultures. *Arch. Intern. Med. 130*, 84-87.

Maiztegui JI, Biegeleisen JZ, Cherry WB, Kass EH (1965) Bacteremia due to gram-negative rods: a clinical, bacteriologic, serologic and immunofluorescent study. *N. Engl. J. Med. 272*, 222-229.

Maki DG (1978) Control of colonization and transmission of pathogenic bacteria in the hospital. *Ann. Intern. Med. 89*, 777-780.

Maki DG (1980) Through a glass darkly, nosocomial pseudoepidemics and pseudobacteremias (Editorial). *Arch. Intern. Med. 140*, 26-28.

Maki DG (1981) Epidemic nosocomial bacteremias. In: Wenzel RP (Ed), *Handbook of Hospital Acquired Infection*, pp 371-512. CRC Press, Boca Raton, FL.

Maki DG (1982) Infections associated with intravascular lines. In: Schwartz M, Remington J (Eds), *Current Clinical Topics in Infectious Disease*, pp 309-363. McGraw-Hill, New York.

Maki DG, Hassemer CA (1981) Endemic rate of fluid contamination and related septicemia in arterial pressure monitoring. *Am. J. Med. 70*, 733-738.

Maki DG, Martin WT (1975) Nationwide epidemic of septicemia caused by contaminated infusion products. IV. Growth of microbial pathogens in fluids for intravenous infusion. *J. Infect. Dis. 131*, 267-272.

Maki DG, Rhame FS, Mackel DC, Bennett JV (1976) Nationwide epidemic of septicemia caused by contaminated intravenous products. I. Epidemiological and clinical features. *Am. J. Med. 60*, 471-485.

Mangurten HH, LeBeau LJ (1977) Diagnosis of neonatal bacteremia by a micro-blood culture technique: a preliminary report. *J. Pediatr. 90*, 990-992.

McCabe WR (1972) Immunization with R mutants of *S. minnesota*. I. Protection against challenge with heterologous gram-negative bacilli. *J. Immunol. 108*, 601-610.

McCabe WR (1973) Serum complement levels in bacteremia due to gram-negative organisms. *N. Engl. J. Med. 288*, 21-23.

McCabe WR (1974) Letter to the editor. *N. Engl. J. Med. 292*, 111.

McCabe WR, Jackson GG (1962a) Gram-negative bacteremia. I. Etiology and ecology. *Arch. Intern. Med. 110*, 83-91.

McCabe WR, Jackson GG (1962b) Gram-negative bacteremia. II. Clinical, laboratory, and therapeutic considerations. *Arch. Intern. Med. 110*, 92-100.

McCabe WR, Kreger BE, Johns M (1972) Type-specific and cross-reactive antibodies in gram-negative bacteremia. *N. Engl. J. Med. 287*, 261-267.

McCabe WR, Greely A, DiGenio T, Johns MA (1973) Humoral immunity to type-specific and cross-reactive antigens of gram-negative bacilli. *J. Infect. Dis. 128*,

S284-S289.

McGowan JE Jr, Garner C, Wilcox C, Finland M (1974) Antibiotic susceptibility of gram-negative bacilli isolated from blood cultures: results of tests with 35 agents and strains from 169 patients at Boston City Hospital during 1972. *Am. J. Med.* *57*, 225-238.

McGowan JE, Barnes MW, Finland M (1975) Bacteremia at Boston City Hospital: occurrence and mortality during 12 selected years (1935-1972), with selected reference to hospital-acquired cases. *J. Infect. Dis. 132*, 316-334.

McGowan JE, Parrot PL, Duty VP (1977) Nosocomial bacteremia: potential for prevention of procedure-related cases. *J. Am. Med. Assoc. 237*, 2727-2729.

McHenry MC, Gaven TL, Hawk WA, Ma C, Berrettoni JN (1975) Gram-negative bacteremia: variable clinical course and useful prognostic factors. *Cleveland Clinic Q. 42*, 1532.

McLaughlin JC, Hamilton P, Scholes JV, Bartlett RC (1983) Clinical laboratory comparison of lysis-centrifugation and BACTEC radiometric blood culture techniques. *J. Clin. Microbiol. 18*, 1027-1031.

Miller RM, Polakavetz SH, Hornick RB, Cowley RA (1973) Analysis of infections acquired by the severely injured patient. *Surg. Gynecol. Obstet. 137*, 7-10.

Mizock B (1984) Septic shock: a metabolic perspective. *Arch. Intern. Med. 144*, 579-585.

Moncrief JA (1978) Topical therapy for control of bacteria in the burn wound. *World J. Surg. 2*, 151-165.

Murphy JF, McDonald FD, Dawson M, Reite A, Turcotte J, Fekety FR Jr (1976) Factors affecting the frequency of infection in renal transplant recipients. *Arch. Intern. Med. 136*, 670-677.

Myerowitz RL, Medeiros AA, O'Brien TF (1971) Recent experience with bacillemia due to gram-negative organisms. *J. Infect. Dis. 124*, 239-246.

National Research Council Division of Medical Sciences, Ad Hoc Committee of the Committee of Trauma (1964) Postoperative wound infections: the influence of ultraviolet irradiation of the operating room and various other factors. *Ann. Surg. 160 (Suppl 2)*, 1-192.

Nelson JD, Richardson J, Shelton S (1965) The significance of bacteremia with exchange transfusions. *J. Pediatr. 66*, 291-295.

Nishijima H, Weil MH, Shubin H, Cavanilles J (1973) Hemodynamic and metabolic studies on shock associated with gram-negative bacteremia. *Medicine (Baltimore) 52*, 287-294.

Northey D, Adess ML, Hartsuck JM, Rhoades ER (1974) Microbial surveillance in a surgical intensive care unit. *Surg. Gynecol. Obstet. 139*, 321-325.

Parsons R, Gregory J, Palmer DL (1983) Salmonella infection of the abdominal aorta. *Rev. Infect. Dis. 5*, 227-231.

Pepe PE, Potkin RT, Reus DH, Hudson LD, Carrico CJ (1982) Clinical predictors of the adult respiratory distress syndrome. *Am. J. Surg. 144*, 124-130.

Phair JP, Bassaris HP, Williams JE, Metzger E (1983) Bacteremic pneumonia due to gram-negative bacilli. *Arch. Intern. Med. 143*, 2147-2149.

Pizzo PA (1984) Empiric therapy and prevention of infection in the immunocompromised host. In: Mandell GL, Douglas RG, Bennett JE (Eds), *Principles and Practices of Infectious Diseases*, pp 1680-1688. John Wiley and Sons, New York.

Pollack M, Charache P, Nieman RE, Jett MP, Reinhardt JA, Harby PH Jr (1972) Factors influencing colonization and antibiotic-resistance patterns of gram negative bacteria in hospital patients. *Lancet 2*, 668-671.

Pruitt BA Jr (1979) The burn patient: II. Later care and complications of thermal injury. *Curr. Probl. Surg. 16*, 1-62.

Raucher HS, Hyatt AC, Barzilai A, Hodes DS (1983) Quantitative blood cultures in the evaluation of Broviac catheter sepsis. In: *Abstracts of the Twenty-Third Interscience Conference on Antimicrobial Agents and Chemotherapy*, p 103. American Society for Microbiology, Washington DC.

Rea WJ, Wyrick WJ Jr (1970) Necrotizing fasciitis. *Ann. Surg. 172*, 957-964.

Rector WG Jr (1981) Fever, shock and chills in gram-negative bacillemia: clinical correlations in 100 cases. *Johns Hopkins Med. J. 149*, 175-178.

Reller LB, Murray PR, MacLowry JD (1982) In: Washington JA, II (Ed), *Cumitech 1A, Blood Cultures II*, pp 1-11. American Society for Microbiology, Washington D.C.

Rhame FS, Root RK, MacLowry JD, Dadisman TA, Bennett JV (1973) Salmonella septicemia from platelet transfusions: study of outbreak traced to hematogenous carrier of *Salmonella cholerae-suis. Ann. Intern. Med. 78*, 633-641.

Rose HD, Babcock JB (1975) Colonization of intensive care unit patients with gram-negative bacilli. *Am. J. Epidemiol. 101*, 495-501.

Rose HD, Schreier J (1968) The effect of hospitalization and antibiotic therapy on gram-negative fecal flora. *Am. J. Med. Sci. 255*, 228-236.

Rubin HM, Blan EB, Michaels RH (1975) Hemophilus and pneumococcal peritonitis in children with the nephrotic syndrome. *Pediatrics 56*, 598-601.

Rubin J, Rogers WA, Taylor HM, Everett ED, Prowant BF, Fruto LV, Nolph KD (1980) Peritonitis during continuous ambulatory peritoneal dialysis. *Ann. Intern. Med. 92*, 7-13.

Sanford JP (1975) Urinary tract symptoms and infection. *Annu. Rev. Med. 26*, 485-498.

Santosham M, Moxon ER (1977) Detection and quantitation of bacteremia in childhood. *J. Pediatr. 91*, 719-721.

Sasaki TM, Welch GW, Herndon DN, Kaplan JZ, Lindberg RB, Pruitt BA (1979) Burn wound manipulation-induced bacteremia. *J. Trauma 19*, 46-48.

Scheckler WE (1977) Septicemia in a community hospital 1970-1973. *J. Am. Med. Assoc. 237*, 1938-1941.

Schimpff SC, Wiernick PH, Block JB (1972a) Rectal abscesses in cancer patients. *Lancet 2*, 844-847.

Schimpff SC, Young VM, Greene WH, Vermeulen GD, Moody MR, Wiernik PH (1972b) Origin of infection in acute nonlymphocytic leukemia: significance of hospital acquisition of potential pathogens. *Ann. Intern. Med. 72*, 707-714.

Schroter GPJ, Hoelscher M, Putnam CW, Porter KA, Hansbrough JF, Starzl TE (1976) Infections complicating orthotopic liver transplantation: with emphasis on graft related septicemia. *Arch. Surg. 111*, 115-122.

Schwartz SN, Dowling JN, Benkovic C (1978) Sources of gram-negative bacilli colonizing the tracheae of intubated patients. *J. Infect. Dis. 138*, 227-231.

Selwyn S, Ellis H (1972) Skin bacteria and skin disinfection reconsidered. *Br. Med. J. 1*, 136-138.

Setia U, Gross PA (1977) Bacteremia in a community hospital: spectrum and mortality. *Arch. Intern. Med. 137*, 1698-1701.

Singer C, Kaplan MH, Armstrong D (1977) Bacteremia and fungemia complicating neoplastic disease: a study of 364 cases. *Am. J. Med. 62*, 731-742.

Smith-Ericksen N, Aasen AO, Gallimore MJ, Amundsen E (1982) Studies of components of the coagulation systems in normal individuals and septic shock patients. *Circ. Res. 9*, 491-497.

Snydman DR, Murray SA, Kornfeld SI, Majka JA, Ellis CA (1982) Total parenteral nutrition-related infections: prospective epidemiologic study using semiquantitative methods. *Am. J. Med. 73*, 695-699.

Spengler RF, Greenough WB III (1978) Hospital costs and mortality attributed to nosocomial bacteremias. *J. Am. Med. Assoc. 240*, 2455-2458.

Spengler RF, Greenough WB III, Stolley PD (1978) A descriptive study of nosocomial bacteremias at the Johns Hopkins Hospital, 1968-1974. *Johns Hopkins Med. J. 142*, 77-84.

Stamenkovic I, Lew PD (1984) Early recognition of potentially fatal necrotizing fasciitis: the use of frozen section biopsy. *N. Engl. J. Med. 310*, 1689-1693.

Stamm WE, Martin SM, Bennett JV (1977) Epidemiology of nosocomial infections due to gram-negative bacilli: aspects relevant to development and use of vaccines. *J. Infect. Dis. 136*, S151-S160.

Steere AC, Rifaat MK, Seligmann EB Jr, Hochstein HD, Friedland G, Dasse P, Wustrack KO, Axnick KJ, Barker LF (1978) Pyrogen reactions associated with the infusion of normal serum albumin (human). *Transfusion 18*, 102-107.

Stevens RM, Teres D, Skilman JJ (1974) Pneumonia in an intensive care unit. *Arch. Intern. Med. 134*, 106-111.

Stone HH, Martin JJ Jr (1972) Synergistic necrotizing cellulitis. *Ann. Surg. 175*, 702-711.

Stratford B, Gallus AS, Matthiesson AM, Dixson S (1968) Alteration of superficial bacterial flora in severely ill patients. *Lancet 1*, 68.

Sullivan NM, Sutter VM, Mims MM, Marsh VH, Finegold SM (1973) Clinical aspects of bacteremia after manipulation of the genitourinary tract. *J. Infect. Dis. 127*, 49-55.

Sullivan RJ, Dowdle WR, Marine WM, Hierholzer JC (1972) Adult pneumonia in a general hospital. Etiology and host risk factors. *Arch. Intern. Med. 129*, 935-942.

Tenney JH, Reller LB, Mirrett S, Wag WL, Weinstein MP (1982) Controlled evaluation of the volume of blood cultured in detection of bacteremia and fungemia. *J. Clin. Microbiol. 15*, 558-561.

Tillotson JR, Finland M (1969) Bacterial colonization and clinical superinfection of the respiratory tract complicating antibiotic treatment of pneumonia. *J. Infect. Dis. 119*, 597-624.

Tonnesen A, Peuler M, Lockwood WR (1976) Cultures of blood drawn by catheters vs. venipuncture. *J. Am. Med. Assoc. 235*, 1877.

Turck M, Stamm W (1981) Nosocomial infection of the urinary tract. *Am. J. Med. 70*, 651-654.

Vas SI (1983) Microbiologic aspects of chronic ambulatory peritoneal dialysis. *Kidney Int. 23*, 83-92.

Veazey JM (1981) Hospital-acquired pneumonia. In: Wenzel RP (Ed), *Handbook of Hospital Acquired Infection*, pp 341-369. CRC Press, Boca Raton, FL.

Waisbren BA (1951) Bacteremia due to gram-negative bacilli other than salmonella: a clinical and therapeutic study. *Arch. Intern. Med. 88*, 467-488.

Wardle N (1982) Acute renal failure in the 1980s: the importance of septic shock and of endotoxemia. *Nephron 30*, 193-200.

Washington JA II (1978) Conventional approaches to blood culture. In: Washington JA II (Ed) *The Detection of Septicemia*, pp 41-81. CRC Press, West Palm Beach.

Weil MH, Shubin H, Biddle M (1964) Shock caused by gram-negative organisms: analysis of 169 cases. *Ann. Intern. Med. 60*, 384-400.

Weinstein L, Goldfield M, Chang TW (1954) Infections occurring during chemotherapy: a study of their frequency, type and predisposing factors. *N. Engl. J. Med. 251*, 247-255.

Weinstein MP, Iannini PB, Stratton CW, Eickhoff TC (1978) Spontaneous bacterial peritonitis: a review of 28 cases with emphasis on improved survival and factors influencing prognosis. *Am. J. Med. 64*, 592-598.

Weinstein MP, Reller LB, Murphy JR, Lichtenstein KA (1983a) The clinical significance of positive blood cultures: a comprehensive analysis of 500 episodes of bacteremia and fungemia in adults. I. Laboratory and epidemiologic observations. *Rev. Infect. Dis. 5*, 35-53.

Weinstein MP, Murphy JR, Reller LB, Lichtenstein KA (1983b) The clinical significance of positive blood cultures: a comprehensive analysis of 500 episodes of bacteremia and fungemia in adults. II. Clinical observations, with special reference to factors influencing prognosis. *Rev. Infect. Dis. 5*, 54-70.

Weinstein WM, Onderdonk AB, Bartlett JG, Louie TJ, Gorbach SL (1975) Antimicrobial studies of experimental intraabdominal sepsis. *J. Infect. Dis. 132*, 282-286.

Welch JP, Donaldson GA (1976) The urgency of diagnosis and surgical treatment of acute suppurative cholangitis. *Am. J. Surg. 131*, 527-532.

White JJ, Wallace CK, Burnett LS (1970) Drug letter: skin disinfection. *Johns Hopkins Med. J. 126*, 169-176.

Wilfert CM, Katz SL (1969) Etiology of bacterial sepsis in nephrotic children 1963-1967. *Pediatrics 42*, 840-843.

Winslow EJ, Loeb HS, Rahimtoola SH, Kamath S, Gunnar RM (1973) Hemodynamic studies and results of therapy in 50 patients with bacteremic shock. *Am. J. Med. 54*, 421-432.

Wolfe MS, Armstrong D, Louria DB, Blevins A (1971) Salmonellosis in patients with neoplastic disease. *Arch. Intern. Med. 128*, 546-554.

Wolff SM, Bennett JV (1974) Gram-negative rod bacteremia. *N. Engl. J. Med. 291*, 733-734.

Young LS, Meyer RD, Armstrong D (1973) *Pseudomonas aeruginosa* vaccine in cancer patients. *Ann. Intern. Med. 79*, 518-527.

Young LS, Stevens P, Ingram J (1975) Functional role of antibody against 'core' glycolipid of Enterobacteriaceae. *J. Clin. Invest. 56*, 850-861.

Ziegler EJ, Douglas H, Sherman JE, Davis CE, Braude AI (1973) Treatment of *E. coli* and Klebsiella bacteremia in agranulocytic animals with antiserum to a UDP-GAL epimerase-deficient mutant. *J. Immunol. 111*, 433-438.

Ziegler EJ, McCutchan JA, Fierer J, Glauser MP, Sadoff JC, Douglas H, Braude AI (1982) Treatment of gram-negative bacteremia and shock with human antiserum to a mutant *Escherichia coli. N. Engl. J. Med. 307*, 1225-1230.

Zieve PD, Haghshenass M, Blanks M, Krevans JR (1966) Vacuolization of the neutrophil: an aid in the diagnosis of septicemia. *Arch. Intern. Med. 118*, 356-357.

Handbook of Endotoxin, Vol. 4: Clinical Aspects of Endotoxin Shock
R.A. Proctor, editor
© Elsevier Science Publishers B.V., 1986

CHAPTER 4

Clinical applications of the Limulus amebocyte lysate test

JAMES H. JORGENSEN

1. BACKGROUND AND MECHANISM OF THE LIMULUS AMEBOCYTE LYSATE–ENDOTOXIN REACTION

The horseshoe crab, *Limulus polyphemus*, is a phylogenetically primitive marine arthropod which has not evolved significantly over the past 200–300 million years (Levin and Bang 1968). It inhabits the Atlantic Ocean from Nova Scotia southward to the Yucatan peninsula. The horseshoe crab has an open circulatory system containing blue hemolymph (due to hemocyanin as the respiratory pigment). The only formed element present in the hemolymph is a cell called the amebocyte (Levin and Bang 1964).

Use of the Limulus amebocyte lysate (LAL) as an in vitro test for endotoxin resulted directly from the important observations made by Bang (1958) at the Woods Hole Oceanographic Institution; he noticed that the horseshoe crab underwent a type of disseminated intravascular coagulation when generalized infection occurred with marine gram-negative bacteria. This initial in vivo observation was later extended by the discovery that clotting of Limulus hemolymph could be produced in vitro by addition of either viable gram-negative bacteria or purified gram-negative endotoxin (Levin and Bang 1964). It was also discovered by the same workers (Levin and Bang 1968) that the amebocyte was the apparent source of all of the factors necessary for hemolymph coagulation. The coagulation components were further localized to the dense granules which packed the cytoplasm of the amebocyte (Murer et al 1975). It was shown that when tissue damage occurred in the crab, the amebocytes aggregated, degranulated, and finally hemolymph coagulation occurred (Dumont et al 1966; Levin and Bang 1964).

A primordial immunologic function appeared to be served by activation of the coagulation pathway of the crab through an alternate mechanism when gram-negative bacteria or their endotoxins were present. In this manner, invading bacteria might be localized and their spread limited by clotting of the hemolymph. The current application of the test as an in vitro assay for endotoxin is largely based on the fact that physical disruption of amebocytes which have been harvested under suitable conditions yields a suspension (the Limulus amebocyte lysate or LAL) containing the coagulation components which may then be activated only by bacterial endotoxin (since key membrane receptors for in vivo activation of the coagulation pathway are removed in lysate preparation; (Levin and Bang 1968). Application of this principle has made the LAL test the most sensitive method available (Rojas-Corona et al 1969) for detection of gram-negative bacterial lipopolysaccharide (LPS, endotoxin). LAL prepared by current purification methods can reliably detect 0.25 ng/ml of purified *Escherichia coli* standard endotoxin. While *L. polyphemus* is the horseshoe crab found along the Atlantic seashore of North America, similar horseshoe crabs, *Tachypleus tridentatis*, are found in the sea of Japan (Nakamura et al 1976), while *Tachypleus gigas* and *Carcinoscorpius rotunda cauda* occur in the Indian Ocean (Mahalanabis et al 1979). It appears that the coagulation systems of the three species are virtually identical, and in fact much of our understanding of the biochemistry of the LAL reaction is based on studies conducted with *T. tridentatis* lysate (Iwanaga et al 1978; Nakamura et al 1976, 1977). Amebocyte lysate of similar reactivity to endotoxin may be prepared from all of these horseshoe crab species.

LAL has been shown to detect either bound (cell-associated) or free endotoxin (Jorgensen and Smith 1974) incorporated in the cell walls of virtually all gram-negative bacteria. However, endotoxins prepared by different extraction procedures, from different species, or even different lots prepared by the same method and from the same species, may vary widely in reactivity (Fleishman and Fowlkes 1982; Weary et al 1980). Certain microorganisms, such as *Neisseria gonorrhoeae* and *Legionella pneumophilia*, are said to produce exceptional amounts of endotoxin (Highsmith et al 1978; Spagna et al 1980). For these reasons, a reference standard endotoxin (RSE) has been prepared by the U.S. Food and Drug Administration (U.S.F.D.A.) as a means for standardization of LAL and rabbit pyrogenicity tests (United States Pharmacopeia XX). By the use of the first RSE, an *E. coli* endotoxin designated EC-2, 1 endotoxin unit was defined as 0.2 ng/ml of EC-2. Future RSE preparations (such as the current EC-5) may then be more easily compared to previous endotoxin and LAL preparations.

It has been demonstrated that, irrespective of source, endotoxin functions as an activator of an inactive proenzyme (MW 150 000 daltons) which is present in LAL (Tai and Liu 1977). Once the enzyme is activated in the presence of suitable divalent cations (such as Ca^{2+}, Mg^{2+}, or Mn^{2+}), it behaves much as a serine protease (amidase) in the hydrolysis of a substrate (coagulogen or clottable protein; MW 19 000–25 000 daltons) which is also present in the LAL. The enzyme is thus similar to thrombin, trypsin, or human coagulation factor Xa. The activated enzyme cleaves the coagulogen molecule at a specific site between the amino acids arginine and lysine (Tai et al 1977). This results in cleavage of the coagulogen molecule into two moieties, the larger of which contains approximately 170 amino acid residues and is referred to as coagulin. The large coagulin molecules then interact with one another to form an aggregated secondary structure. The secondary structure forms a polymerized gel with both hydrophilic and hydrophobic bonds between the large peptide strands (Tai et al 1977). Electron microscopy of the final gel has revealed large aggregated strands whose structure resembles a single alpha-helix (Holme and Solum 1973). The smaller peptide fragments (approximately 45 amino acid residues), which are released from the clottable protein during this process, remain soluble and do not appear to contribute to the integrity of the clot (Tai et al 1977). The visible end result of this cascade of reactions is most often seen as an opaque gel or clot. In the endotoxin assay, when the clot is formed in a reaction tube, this is recognized by failure of the contents to flow when the test tube is inverted 180°. Less intense reactions have been described, which manifest as an increase in viscosity, appearing as starch-like granules which adhere to the wall of a reaction tube, often accompanied by increase in opacity (Jorgensen and Smith 1973a). The subjectivity of interpreting endpoints of less than complete gelation has been a point for criticism of this method of performing the test.

Alternative methods for detecting endotoxin by LAL include observing the increase in optical density which appears to occur as a result of the clottable protein polymerization (Jorgensen and Alexander 1981; Levin and Bang 1968; Teller and Kelly 1979); direct measurement of the action of the enzyme on a synthetic chromogenic substrate (Iwanaga et al 1978; Nakamura et al 1977); and an immunologic test (rocket immunoelectrophoresis) which measures the loss or conversion of coagulogen from lysate when the proenzyme is activated by endotoxin by use of antibody directed against the intact coagulogen molecule (Back 1983). With any of these methods, the slowest and rate-limiting step in the reaction is the time required for activation of the proenzyme by endotoxin (Young et al 1972). Most LAL test methods include periods of 1 hour or more with a minimum of 5–10 minutes required for the earliest signs of the initial proenzyme activation to be detectable. The majority of clinical studies of the LAL assay have employed the traditional gelation end-point method of performing the test. Only recently, chromogenic substrate or turbidometric methods have been applied in clinical studies of the test.

The specificity of the LAL reaction for LPS endotoxin has been questioned by several investigators. Thrombin, thromboplastin, ribonucleases, and certain synthetic polynucleotides were said to cause gelation of LAL by Elin and Wolff (1973). However, later studies (Pearson and Weary 1980; Yin 1975) suggested that the LAL reactivity could have been due to inadvertent endotoxin contamination of the preparations originally tested by Elin and Wolff. Certain components of gram-positive bacterial cell walls or exotoxins of some streptococci were shown to react at very high concentrations with LAL (Pearson and Weary 1980). Likewise certain synthetic dextrans, or simple polysaccharides, again in high concentrations, appeared to provide positive LAL reactions as well as pyrogenic reaction in rabbits (Mikami et al 1982; Suzuki et al 1977). Lastly, Platica et al (1978) demonstrated gelation of LAL by the thio compounds dithiothreitol and dithioerythritol using a spectrophotometric LAL test. The likelihood of any of these materials causing false-positive LAL reactions from human diagnostic specimens of pharmaceutical products, however, is remote since these materials would not be expected to be found under physiologic conditions, or to be present in the high concentrations reported to cause positive LAL reactions (Pearson and Weary 1980).

Certain substances have been described which either enhance or inhibit the LAL reaction with endotoxin. Calcium gluconate was shown to enhance the LAL reaction and thus falsely elevate endotoxin levels in a contaminated pharmaceutical preparation (Steindler et al 1981). Because the LAL gelation is mediated by a trypsin-like enzyme reaction, substances or conditions which denature protein or inhibit enzyme activity may result in false-negative LAL reactions. Unfavorable pH, sodium chloride, heparin, certain penicillins, urea, and dimethyl sulfoxide have all been reported to retard or inhibit the activation of proenzyme present in LAL (Levin et al 1970b; Liang and Liu 1982; McCullough and Scolnick 1976; Schleef et al 1979, 1982; Sullivan and Watson 1975).

2. LIMULUS AMEBOCYTE LYSATE TESTING IN THE PHARMACEUTICAL INDUSTRY

Several human clinical applications of the LAL test have been shown to be very promising and shall be discussed in this chapter. However, by far the largest user of the LAL test has been the pharmaceutical industry, which now performs hundreds of thousands of tests annually on a variety of large and small volume parenteral fluids, biologicals, and medical devices (Cooper and Neely 1980), and is amendable to the possibility of automation (Jorgensen and Alexander 1981). The LAL test is now commonly performed for both in-process and final-release testing of many injectable fluids and invasive medical devices produced by the U.S. pharmaceutical industry. This continues to be the only U.S.F.D.A.-licensed application of the LAL test.

3. CLINICAL APPLICATIONS OF THE LIMULUS AMEBOCYTE LYSATE TEST

3.1. Detection of endotoxemia

The first clinical application of the LAL assay, detection of endotoxemia during septicemia, has unfortunately been clouded by controversy regarding its true diagnostic usefulness. The presence of 'inhibitors' of the LAL test in human blood has made it difficult to reliably detect endotoxin in blood and has required some preparative procedure to inactivate the inhibitors before testing with LAL. The utility of several of these methods will be discussed in this chapter, but none of them has yet proved to be universally acceptable.

The exact identity of the inhibitor(s) has not been conclusively determined, but it appears to be an alpha globulin which probably serves a host defense mechanism by disaggregating or reversibly binding free endotoxin (Johnson et al 1977). Early studies also suggested that endotoxin was trapped in the process of clotting of human blood (Levin et al 1970b; Reinhold and Fine 1971) and was thus not fully recoverable when LAL tests were performed on serum. Virtually all clinical studies of endotoxemia have been performed on plasma, generally using heparin as the anticoagulant which least interferes with the LAL reaction (Levin et al 1970b).

Results of LAL tests on the blood of patients suspected of gram-negative sepsis have generally resulted in rather poor correlation with conventional

TABLE 1 *Summary of selected studies using the LAL assay for detection of endotoxemia*

Plasma treatment method	No. of patients or samples examined	Patients (samples) with positive LAL tests		LAL results compared to:	Gram-negative bacteremia in patients with positive LAL tests
		No.	Rate (%)		
Chloroform extraction	First description on spiked blood	Endotoxin recovered from spiked plasma			
Chloroform extraction	98	17	17	Blood cultures and clinical findings	29% (5/17)
pH Shift	First description, method of quantitation	Quantitative recovery of endotoxin from spiked plasma		–	–
Chloroform extraction	281	39	14	Blood cultures, hypotension, death, blood coagulopathies, and complement consumption	51% (20/39)
pH Shift	24	24	100	Various clinical features	N.D.
Chloroform extraction	10	9	90	Blood cultures, bubo aspirates, and serology for plague	22% (2/9)
pH Shift	83	2	2.5	Blood and miscellaneous cultures	50% (1/2)
Chloroform extraction	139	54	39	Quantitative blood cultures and miscellaneous cultures	52% (28/54)
Platelet and chloroform extractions	54	18	33	Blood and miscellaneous cultures	69% (11/16)
Chloroform extraction	46	33% of tests (79/240)		Localized and generalized infections	12% of tests (2/17)
pH Shift	89	2	2	Blood cultures of pediatric cancer patients	100% (2/2)
Chloroform extraction	18	9	50	Blood and miscellaneous cultures on severely malnourished children	25% (3/12)

Positive LAL test			Conclusions	Reference
Patients with gram-negative bacteremia	Patients with other sites of gram-negative infection	Patients without gram-negative infections		
–	–	–	First description of LAL test on human blood	Levin et al (1970)
71% (12/17)	100% (5/5)	0	Sensitive and reliable for endotoxemia	Levin et al (1970b)
–	–	–	Alternative method for removal of inhibitors, also allows quantitation	Reinhold and Fine (1971)
59% (20/34)	89% (16/18)	0	Endotoxemia correlated well with pathophysiologic alterations attributed to gram-negative sepsis	Levin et al (1972)
N.D.	N.D.	50% (12/24)	Endotoxemia occurs from a variety of septic and nonseptic physiologic disorders	Caridis et al (1972)
100% (2/2)	90% (9/10)	0	Endotoxemia demonstrated in plague	Butler et al (1973)
6% (1/16)	5% (2/39)	N.D.	LAL test of dubious diagnostic utility	Martinez-G et al (1973)
43% (28/65)	35% (18/52)	36% (8/22)	Doubtful diagnostic value of LAL test; no correlation with density of gram-negative bacteria in blood	Stumacher et al (1973)
82% (9/11)	100% (7/7)	0 (0/36)	Extraction of platelets increases accuracy for detection of endotoxemia	Das et al (1973)
100% (2/2)	72% of tests (46/64)	15% of tests (17/117)	Endotoxemia may occur without bacteremia	Fossard et al (1974)
7% (2/30)	0	0	LAL test not clinically useful for predicting gram-negative infections	Feldman and Pearson (1974)
100% (3/3)	40% (6/15)	6% (1/18)	LAL useful to predict gram-negative infections in severely malnourished children	Oberle et al (1974)

TABLE 1　*cont'd*

Plasma treatment method	No. of patients or samples examined	Patients (samples) with positive LAL tests		LAL results compared to:	Gram-negative bacteremia in patients with positive LAL tests
		No.	Rate (%)		
Chloroform extraction	93	N.D.		Blood cultures, fibrin degradation products, platelet and WBC counts on pediatric surgical patients	N.D.
Platelet and chloroform extractions	111	20% of tests (48/237)		Blood cultures and numerous other clinical and laboratory data	100% of tests (8/8)
Chloroform extraction	68	35	52	Opsonizing antibody titers, WBCs and platelets in bacteremic patients	N.D.
Chloroform extraction	10	3	30	Quantitative blood cultures and clinical features of bacteremic plague patients	100% (3/3)
Chloroform extraction	21	0	0	Quantitative blood cultures and blood coagulation factors of typhoid fever patients	0
Chloroform extraction (*Tachypleus* lysate used)	108	16	33	Blood cultures	89% (34/38)
Heating at 65°C for 15-20 min.	115	N.D.		Blood cultures and clinical findings in infants	15% (8/55)
Heating at 100°C for 10 min (spectrophotometric LAL assay)	170	N.D.		Blood cultures, WBCs, liver function	30% (36/119)

N.D. = not determined or not stated.

Positive LAL test			Conclusions	Reference
Patients with gram-negative bacteremia	Patients with other sites of gram-negative infection	Patients without gram-negative infections		
64% (60/93)	N.D.	10% (4/41)	Numerous false-positive and false-negative LAL tests compromise usefulness	Rowe et al (1975)
35% of tests (8/23)	N.D.	15% of tests (36/237)	Potencies of different LAL preparations vary widely; no diagnostic usefulness	Elin et al (1975)
52% (35/68)	N.D.	N.D.	High opsonic titers may reduce LAL test reliability	Young (1975)
60% (3/5)	0 (0/5)	None tested	LAL test positive if blood bacterial count $> 10^2$/ml	Butler et al (1976)
0(0/13)	0 (0/7)	None tested	False-negative LAL test due to low level of bacteremia $< 9 \times 10^2$/ml	Butler et al (1978)
85% (34/40)	N.D.	7% (4/61)	*Tachypleus* lysate useful for detection of endotoxemia	Usawattanakul et al (1979)
88% (7/8)	N.D.	28% (47/167)	LAL test not useful diagnostically during first week of life	Scheifele et al (1981)
97% (35/36)	N.D.	62% (83/134)	Spectrophotometric assay sensitive for diagnosis of gram-negative septicemia; heat-labile factors from WBCs also cause positive LAL tests	Fink et al (1981)

methods of diagnosis such as blood cultures for documentation of gram-negative bacteremia. Such inconsistencies have raised theoretical questions which are difficult to answer conclusively, such as the possibility of bacteremia without endotoxemia, or endotoxemia without demonstrable bacteremia. This section of the discussion will attempt to present the finding of these various studies in an effort to clarify this aspect of the LAL assay.

Table 1 describes the results of eighteen different studies involving more than 1700 patients in which the LAL test was used to diagnose gram-negative septicemia. Until recently, only three methods had been described for treatment of plasma to alleviate the inhibitors of the LAL reaction. These methods include chloroform extraction (2–4 hours with vigorous mixing) to denature the inhibitors, a pH-shift technique with acetic acid and phosphate buffer (to precipitate inhibitors by acid, then restore neutral pH for optimum LAL reactivity), or destruction of the inhibitors by heating in a boiling water bath. Studies using the pH shift method of Reinhold and Fine (1971) have been fewer in number and generally reflect a lower rate of success than those with other plasma treatment methods. The study by Caridis et al (1972) described endotoxemia in patients with a number of infectious disorders such as perforated appendix, pneumonia, perforated sigmoid colon, and other conditions which might reasonably represent generalized gram-negative sepsis. However, other patients were included with a variety of non-septic conditions such as heatstroke, cirrhosis, leukemia, burns, and acute pancreatitis; the authors argued that these patients had circulating endotoxins because of altered permeability of the intestinal wall and subsequent leakage of endotoxins from the normal gram-negative intestinal microflora. The authors attribute the occurrence or continuance of the endotoxemia to malfunction of the reticuloendothelial system. Similarly, Cuevas and Fine (1972) showed uptake of endotoxin from the gut in experimental animals with either septic or chemically induced peritonitis. Jacob et al (1977) demonstrated endotoxin by using the LAL assay in portal venous blood of patients undergoing abdominal surgery. They reasoned that portal endotoxemia was a normal event and that generalized endotoxemia was prevented by the cleansing action of the Kupffer cells of the liver in healthy individuals. Similarly, Wilkinson et al (1974), Cooperstock et al (1975), and Tarao et al (1977) described endotoxemia by LAL testing in a variety of patients whose underlying difficulty centered on cirrhosis or severe hepatic failure. Walker (1978) summarized this line of reasoning by stating that the general immunocompetency of the host determines the occurrence and pos-

136

sibly the persistence of endotoxemia which arises from a gastrointestinal source. Likewise, Jones and Roe (1979) hypothesized that the burn wounds of their patients which were heavily colonized with gram-negative bacteria served as a pool of endotoxin which was absorbed into the circulation in a manner analogous to altered gut permeability.

The hypothesis that endotoxemia is a final common pathway in shock (Ravin et al 1960; Schweinburg and Fine 1960) due to a variety of sources did not cause the authors of the early LAL endotoxemia studies to question endotoxemia without concomitant bacteremia. Caridis et al (1972) suggested that demonstration of endotoxemia by the LAL test in this manner was a grave prognostic sign which required immediate heroic efforts to wash out or to detoxify the endotoxins which were detected. Their report was therefore one of several which found no contraindication in the concept of endotoxemia due to a variety of sources without the simultaneous presence of viable bacteria in the blood, as evidenced by negative blood cultures. Similar theories were advanced by Fossard et al (1974), who used the alternative chloroform extraction method of plasma treatment.

Subsequent studies using the pH-shift method of plasma treatment have seriously questioned the diagnostic usefulness of the LAL test. Martinez-G et al (1973) found that only one of 16 patients with documented gram-negative bacteremia had a positive simultaneous LAL assay. Likewise, Feldman and Pearson (1974) found only 7% (2/30) LAL positivity in the blood of pediatric cancer patients with gram-negative bacteremia. Therefore, both studies concluded that the LAL test by the pH-shift method lacked sensitivity, and therefore diagnostic usefulness. Neither study, however, found false-positive tests, that is, endotoxemia without bacteremia, to be a problem.

The investigations performed by Levin et al (1970a, 1972) represent carefully controlled studies of various physical and laboratory findings associated with gram-negative septicemia. In these studies, the LAL assay was described as a useful means of documenting one phase of sepsis, that is, endotoxemia. The concepts of bacteremia without endotoxemia and endotoxemia without bacteremia were described in a plausible manner as representing two separate, but related phases of the same episode. Distinct correlations were drawn between endotoxemia and hypotension and eventual death of the patient (Levin et al 1972). Likewise, a number of the nonbacteremic, but LAL-positive patients were shown to have documented gram-negative infections at some site which might be expected to lead to transient bacteremia or endotoxemia. However, in two subsequent studies in which chloroform extraction was used, patients who had indwelling urinary catheters (Garibaldi et al 1973) or who underwent cystoscopy or transurethral resection (Robinson et al 1975) had apparent endotoxemia, but not often gram-negative bacteremia. While genitourinary infections or manipulations may be expected to give rise to transient bacteremia, positive blood cultures were rarely encountered in either study. The lack of agreement between the LAL test and the traditional method of blood culture was disquieting from the aspect of diagnostic usefulness. Few clinicians could be expected to fully appreciate the meaning of endotoxemia with negative blood cultures.

An extensive study reported by Stumacher et al (1973) cast serious doubt on any diagnostic usefulness of LAL test results. Their study of 139 patients using the chloroform extraction procedure failed to demonstrate any correlation with the numbers of circulating gram-negative bacteria as determined by quantitative blood cultures. Only 43% of their bacteremic patients had concomitantly positive LAL tests. Likewise, only 52% of patients with endotoxemia documented by the LAL test also had positive blood cultures for gram-negative bacilli. Even if the possibility of localized gram-negative infections of other body sites were considered, only 35% of patients demonstrated positive LAL tests. Conversely, 36% of patients without a likely source of gram-negative infection had apparent endotoxin detected in their blood. Very similar findings were reported in the study by Rowe et al (1975), in which only 64% of bacteremic patients had positive LAL tests, while 10% of uninfected patients had unexplained endotoxemia. These data seriously challenged use of the LAL test for diagnosis of systemic gram-negative infections.

Das et al (1973) described a slight modification of the LAL test which incorporated extraction of platelets, or instead use of platelet-rich plasma for extraction, to detect endotoxin which was absorbed onto the surface of platelets. This improvement of the chloroform extraction technique was said to markedly reduce false-negative LAL tests by increasing the total recovery of endotoxin from blood. When this modification was used, 82% of bacteremic patients were positive by LAL testing and 69% of LAL-positive patients had documented bacteremia. No false-positives were encountered among 36 patients without serious gram-negative infections.

A possible explanation for the controversial and conflicting results obtained when LAL tests were performed on blood was advanced by Elin et al (1975b), who found varying reactivities of different lysate preparations when simultaneous assays of the same positive human plasma samples were performed. However, these authors chose to interpret their results somewhat differently than previous workers. While strong statistical associations were found between endotoxemia and bacteremia (100% of bacteremic patients also had endotoxemia), the fact that only 35% of positive LAL tests were accompanied by bacteremia caused them to conclude that there was no diagnostic usefulness of the LAL test.

Studies by Oberle et al (1974) and investigations involving patients with plague (Butler et al 1973, 1976) demonstrated that the intensity of the bacteremia was strongly predictive of LAL test results. In patients with plague, positive LAL tests were seen only when the density of bacteria in the blood exceeded 10^2/ml (Butler et al 1976). This assumption was similarly applied to explain the negative results obtained when the blood of patients with typhoid fever was examined by the same authors (Butler et al 1978). The lack of apparent endotoxemia in typhoid patients was ascribed to the low level of bacteremia ($< 9 \times 10^2$/ml) documented in the study. Those studies which determined the actual numbers of viable bacteria in blood (Butler et al 1976, 1978; Stumacher et al 1973) and studies with other fluids (Jorgensen et al 1973; Jorgensen and Smith 1974) strongly suggested that the threshold of detectability of bacteria on the basis of the bound endotoxin of the bacterial cells themselves was in the order of 10^2 bacteria/ml or more. Thus, bacteremias of lesser intensity could easily be the explanation for the negative LAL tests on samples which yielded positive blood cultures.

Comments on or documentation of antibiotic therapy as an explanation for endotoxemia without bacteremia was rarely advanced in the preceding studies. However, the theory of rapid lysis of spirochetes during antibiotic therapy causing endotoxemia, and thus the Jarisch-Herxheimer reaction, was documented in syphilis (Gelfand et al 1976) and borreliosis (Galloway et al 1977) by the use of the LAL test. A similar phenomenon was proposed for rickettsial diseases when Fumarola et al (1980) demonstrated LAL activity in several species of *Rickettsia*.

The third of the plasma treatment methods, heat denaturation of the inhibitors, did not seem to be a solution to the previous criticisms of the LAL test. Scheifele et al (1981) concluded that endotoxemia was so prevalent among infants less than one week of age (33%) that the test lacked any diagnostic usefulness. Similarly, all 38 neonates examined by Fink et al (1981) demonstrated apparent, but unexplained endotoxemia as shown by a spectrophotometric version of the LAL test. Fink et al (1981) also found a high rate of LAL positivity using heat treatment of plasma in parturients and cirrhotics whom they examined. Apparently, there is some association of mild heat treatment of plasma with a higher proportion of false-positive, but not false-negative tests. However, DuBose et al (1978) also demonstrated apparent false-positive LAL tests in the plasma of several common research animals and almost one-third of normal human plasma samples examined following the standard chloroform extraction technique. An alternative procedure of heating plasma at a higher temperature (95 °C for 10 min) was used by Goldstein et al (1976), who found only 1 positive test in 55 blood samples from neonates. Both Gaffin et al (1979) and McLeod and Katz (1981) found the mild heat treatment (65–70 °C) to be simple and expedient for denaturation of plasma inhibitors. However, it is not possible to determine whether these authors (Gaffin et al 1979; McLeod and Katz 1981) experienced false-positive LAL reactions when testing heated plasma.

The study by Usawattanakul et al (1979) was the first to report on use of *Tachypleus* lysate for detection of human endotoxemia. With the traditional chloroform extraction method very good results (85% of bacteremic patients were LAL-positive, 89% of endotoxemic patients were also bacteremic) were obtained with few false-positive results (7%). Further studies will be required to determine whether *Tachypleus* lysate proenzyme is less susceptible to the inhibitors in human plasma.

DuBose et al (1980) provided the first comparative evaluation of the efficacy of the various plasma treatment techniques, although all results were based upon spiked normal human blood samples and not on specimens from ill patients. However, it appears that the dilution and heating method of Cooperstock et al (1975) and a method of plasma separation to eliminate inhibitors by gel filtration (Hollander and Harding 1976) provided somewhat greater recovery of endotoxin than was possible by the traditional chloroform extraction procedure. While chloroform extraction was not the most sensitive of the study methods, it was found to be more reliable than the pH-shift method and allowed recovery of approximately 1/50 the amount of spiked endotoxin in the same samples.

Several other modifications of the LAL test for use on blood have been described in recent years. Two of these are spectrophotometric techniques (Hollander and Harding 1976; Nandan et al 1977), which claim much greater sensitivity for detection of the end result of the LAL reaction. It is not clear whether this increased sensitivity will improve the clinical usefulness of the test, however. The use of a synthetic chromogenic substrate to detect activation of the LAL proenzyme has been described for use on blood by several authors (Takagi et al 1981; Tamura et al 1982; Webster 1980). Still newer plasma treatment methods were also incorporated into two of these methods. Takagi et al (1981) described an alkali treatment with NaOH-KCl which proved very selective for removal of the LAL inhibitors (more so than chloroform) in spiked blood samples. Tamura et al (1982) found that treatment of platelet-rich plasma or serum by 1.25% perchloric acid (PCA) completely eliminated nonspecific amidase activity on the synthetic chromogenic substrate and allowed complete recovery of added endotoxin. Thus, use of either PCA or NaOH-KCl, coupled with lysate which incorporates a synthetic chromogenic substrate, may prove to be the ultimate solution to some of the false-positive and false-negative LAL reactions encountered in the clinical studies reported in this section. Certain of these latter studies also challenge the prior concept that serum is not a suitable fluid for detection of blood-borne endotoxin. Webster (1980) did not encounter significant endotoxin trapping when plasma was allowed to clot, forming serum. Indeed, Tamura et al (1982) described excellent recovery of added endotoxin by the PCA treatment irrespective of whether or not anticoagulant was present.

In summary, certain of the more recently described plasma extraction methods or modified lysate preparations may in time rectify the previous shortcomings of the LAL assay for diagnosis of gram-negative septicemia. However, in spite of eloquent arguments to describe the dissociation of bacteremia and endotoxemia (Cuevas and Fine 1972; Levin et al 1972; Sibbald 1976), the diagnostic application of LAL test data from blood has not been proven in carefully conducted studies. The LAL assay suffers in part from lack of another equally sensitive, but different test for endotoxin to which LAL results might be compared for ultimate clarification of these conflicting data. However, until such a test is described or until newer LAL methodologies can resolve such discrepancies, use of the LAL test on blood cannot be viewed with the enthusiasm warranted by the successful uses of the test for diagnosis of gram-negative meningitis, keratitis, bacteriuria, or its unquestioned success in the pharmaceutical industry.

3.2. Rapid diagnosis of gram-negative bacterial meningitis

The use of the LAL test as a means for rapid diagnosis of gram-negative bacterial meningitis is probably the most successful clinical application of the test to date. At least eight separate studies have carefully evaluated LAL test results on cerebrospinal fluid with results of CSF cultures, gram stains, and other laboratory and clinical findings. Table 2 summarizes the results of the eight studies (Berman et al 1976; Butler et al 1976; Dyson and Cassady 1976; Jorgensen and Lee 1978; McCracken and Sarff 1976; Nachum et al 1973; Ross et al 1975; Tuazon et al 1977), which have shown encouraging results on almost 1300 patients with culture-documented meningitis. This application of the LAL test was first described in 1973 (Nachum et al 1973) in a study involving 112 patients suspected of acute bacterial meningitis. The LAL test was positive on all initial CSF specimens from 38 patients with culture-proven gram-negative bacterial meningitis, while negative tests were seen in all 74 patients with gram-positive bacterial, tuberculous, or aseptic meningitis, as well as in patients without meningitis. Patients with meningitis in this study demonstrated growth of the microorganisms most commonly associated with acute bacterial meningitis, that is, *Haemophilus influenzae*, *Neisseria meningitidis*, and *Streptococcus pneumoniae*. While all (38) of the initial CSF specimens from patients with gram-negative meningitis were rapidly positive by LAL assay (usually within 15 minutes), only 66% of the CSF gram-stained smears revealed bacteria (Nachum et al 1973). Thus, the LAL test provided a rapid, objective (nonmicroscopic) means for recognition of bacterial meningitis which clearly surpassed one of the most traditional tests

TABLE 2 *Summary of eight studies using the LAL test for diagnosis of bacterial meningitis*

No. of patients examined	Positive LAL tests		Reference
	Bacterial meningitis	Gram-negative bacterial meningitis	
112	88% (38/43)	100% (38/38)	Nachum et al (1973)
335	73% (37/51)	97% (37/38)	Ross et al (1975)
232	80% (86/107)	100% (86/86)	Berman et al (1976)
3	100% (3/3)	100% (3/3)	Butler et al (1976)
145	60% (6/10)	100% (6/6)	Dyson and Cassady (1976)
84	64% (60/94)	71% (60/84)	McCracken and Sarff (1976)
81	57% (8/14)	100% (8/8)	Tuazon et al (1977)
305	82% (61/74)	100% (61/61)	Jorgensen and Lee (1978)

for this purpose, gram-stain. Nachum et al (1973) also demonstrated the utility of a non-growth-dependent test such as the LAL in the accurate diagnosis of five cases of partly antibiotic-treated meningitis. This initial study, and those which have followed it, have demonstrated almost 100% detection of gram-negative meningeal pathogens, including *H. influenzae, N. meningitidis,* various Enterobacteriaceae, and pseudomonads. However, a major shortcoming demonstrated by all of the studies is the inability of the LAL test to detect gram-positive bacteria, most notably *S. pneumoniae,* Group B Streptococcus and *Staphylococcus aureus.*

When data from all eight published studies regarding meningitis are considered together, well over 90% of patients with documented gram-negative meningitis were rapidly and objectively detected by the LAL test (Table 2). This compares to a detection rate by the CSF gram-stain of only about 70% in these same studies. Only the study of neonatal meningitis reported by McCracken and Sarff (1976) has indicated a lower rate of positivity of the LAL test on CSF. A methodologic difference in the latter study was that CSF was tested which had been frozen for varying periods before the LAL assay instead of being tested immediately.

The study by Berman et al (1976) employed the LAL test to quantify the amount of endotoxin present in the CSF as a possible prognostic indicator of eventual clinical outcome. However, CSF endotoxin concentrations varied widely (4–2000 ng/ml) and did not correlate with at least initial clinical or laboratory variables. Moreover, CSF endotoxin levels were found to decrease quickly with appropriate antibiotic therapy, to undetectable levels within 5 days. In no case were bacteria recovered from CSF samples that were LAL test-negative, although endotoxin was found to persist in the CSF after viable organisms ceased to be isolated (Berman et al 1976). Jorgensen and Lee (1978) found that CSF endotoxin could persist (with sterile cultures) for as long as 9 days after initiation of appropriate antibiotic therapy, and was most likely to occur with *H. influenzae.* However, when the LAL assay of CSF continued to be positive in patients with meningitis due to other gram-negative species, the persistence of endotoxin was associated with a failure to sterilize the CSF, as manifested by continued positive cultures as well as CSF endotoxin (Jorgensen and Lee 1978).

The main predictive value of the LAL assay of CSF is its low percentage of false-positive reactions (less than 2% if performed carefully). Thus, a patient with a positive LAL test of scrupulously collected CSF should be presumed to have a gram-negative bacterium of some species in his CSF.

However, a negative test cannot preclude the possibility of meningitis, especially when due to a gram-positive microorganism. We have also encountered 5 patients over the past 5 years in which the LAL test appeared to be falsely negative, as compared to CSF culture (previously unpublished findings). All of these patients had either brain or parameningeal abscesses which were associated with extremely low numbers (< 20/ml) of gram-negative bacteria found in the CSF. Thus, while these patients did not appear to have meningitis per se, they did have gram-negative bacilli recovered from the CSF in spite of negative LAL tests.

The LAL assay has thus been established as a useful rapid test for diagnosis of gram-negative meningitis. This application of the test is primarily successful because the problem of plasma protein inhibitors is avoided by testing the CSF directly (which does not appear to contain any inhibitors) and the number of gram-negative bacteria which are commonly found in the CSF during meningitis (10^3–10^8/ml; Feldman 1976) easily exceeds the threshold sensitivity of the LAL test (approximately 10^2/ml). However, its inability to detect gram-positive bacteria makes it impossible to rely on it as the sole laboratory criterion for establishing a diagnosis of bacterial meningitis. LAL test results may be very effectively combined with results of simultaneous counterimmunoelectrophoresis (CIE) or particle agglutination tests for bacterial antigens in CSF, which may detect organisms missed by LAL, for example, *S. pneumoniae* and Group B Streptococcus, or which may clarify the species identity of gram-negative bacteria associated with a positive LAL test, for instance, *H. influenzae* or *N. meningitidis*. Neither of these methods should be performed in lieu of a smear of the CSF which might reveal Staphylococci, *Listeria* species, or other microorganisms which are not routinely detectable by LAL, CIE, or particle agglutination. A final shortcoming of the LAL test for diagnosis of meningitis is the fact that it has not been approved by the U.S.F.D.A. for in vitro diagnosis of meningitis or any other human infectious process.

3.3. Detection of gram-negative bacteriuria

Diagnosis of urinary tract infection traditionally involves cultivation of viable bacteria from a urine specimen which may require 24–48 hours incubation for results. Several more rapid techniques have been proposed to detect bacterial products in urine (Neter 1965), provide a simple means for microscopic visualization of bacteria in urine (Brody et al 1968), or rapidly recognize microbial growth by photometric detection of turbidity in broth cultures (Pezzlo et al 1982). Several studies have demonstrated that the LAL test may also be used for this purpose (Table 3). Jorgensen et al (1973) and Jorgensen and Jones (1975) have shown that gram-negative bacteria in numbers which represent significant bacteriuria ($\geq 10^5$/ml of urine) may be reliably detected by using diluted urine and gelation LAL testing. Dilution of urine specimens was found to be necessary to differentiate between significant bacteriuria and lower numbers of inadvertent contaminants from the distal urethra or vagina encountered during normal collection procedures. These two studies of more than 1200 patients showed that 99% of urines indicative of significant gram-negative bacteriuria were detected by the LAL test. However, as previously discussed, gram-positive bacteria and fungi were not detectable using LAL. Thus, only 87% of all urines demonstrating significant bacteriuria (both gram-positive and gram-negative) were detected by the LAL test (Table 3). The overall ability of the test to correctly distinguish between the presence or absence of significant bacteriuria was 95% in these two studies (Jorgensen et al 1973; Jorgensen and Jones 1975).

In a later study by Nachum and Shanbrom (1981), a different method of LAL quantitation was used for recognition of significant bacteriuria, namely the time till LAL gelation when undiluted urine was tested. All urines which contained more than 10^5/ml gram-negative bacteria yielded positive LAL tests within 15 minutes of incubation. By the use of this method of quantitation it was possible to differentiate 96% of urines correctly, as either containing less than 10^5 or 10^5 or more bacteria/ml of urine (Table 3). This is in spite of the fact that both sodium chloride and urea have been found to inhibit the LAL reaction (Schleef et al 1979, 1982) and would be expected in relatively high concentrations in human urine.

The most recent attempt to apply the LAL test for detection of bacteriuria made use of automated, spectrophotometric interpretation of LAL reactions (Jorgensen and Alexander 1982). An Abbott MS-2 automated instrument was used to detect kinetic changes in optical density which occurred when urine samples (diluted 1:2000) containing 10^5/ml or more gram-negative bacteria were incubated with standard LAL in special MS-2 cuvettes. The total LAL test time used in this study was 20 minutes, with positive urines usually detected in the first 5–10 minutes of the test. The accuracy of the automated method was comparable to the three manual gelation studies previously cited (Table 3). Advantages of the automated LAL test for this purpose included increased speed of detection of positives, and a totally objective indication

TABLE 3 *Summary of four studies using the LAL test for detection of significant gram-negative bacteriuria*

LAL test method	No. of patients or urines tested	Positive LAL tests		Distinction between urines with $< 10^5$/ml or $\geq 10^5$/ml	Reference
		$\geq 10^5$ organism/ml	$\geq 10^5$ gram-negative bacteria/ml		
Gelation testing of diluted urine	209	92% (23/25)	100% (23/23)	99% (207/209)	Jorgensen and Smith (1973)
Gelation testing of diluted urine	1077	86% (175/203)	99% (159/161)	94% (1016/1077)	Jorgensen and Jones (1975)
Time to gelation	190	84% (36/43)	100% (36/36)	93% (177/190)	Nachum and Shanbrom (1981)
Automated turbidometric	580	85% (78/92)	94% (78/83)	93% (537/580)	Jorgensen and Alexander (1982)

of a positive test (determined by kinetic changes in optical density indicated by the instrument which exceeded a pre-established threshold value).

The usefulness of the LAL test for detection of bacteriuria lies in the predominance of gram-negative species which cause urinary tract infections: almost 90% of all infections are due to gram-negatives. The almost total detection of gram-negatives, but lack of recognition of gram-positives results in an overall rate of detection by LAL very similar to that of current growth-screening methods (Pezzlo et al 1982). Thus, the LAL test might be a useful technique to rapidly assess the likelihood of bacteriuria in high-risk patients, but certainly could not be considered to be as definitive as performance of a quantitative bacterial culture.

3.4. Presumptive diagnosis of gonorrhea

One of the latest clinical applications of LAL has been for diagnosis of gonococcal urethritis and cervicitis which was first described by Spagna et al (1979) in a study of gonococcal urethritis in men. This initial work demonstrated that use of a large dilution (1:200) of otherwise untreated urethral exudate prior to testing allowed distinction between gonococcal and nongonococcal urethritis when tested by LAL. The dilution of the sample was found to be necessary to prevent false-positive LAL tests from gram-negative flora which normally inhabit the distal urethra of healthy individuals; it also seemed to exclude positive LAL tests due to infections caused by presumed gram-negative microorganisms such as *Chlamydia* and *Mycoplasma*. Support for this hypothesis was provided in a subsequent communication (Spagna et al 1980), in which endotoxin from gonococci appeared to be either more reactive with LAL, or possibly more abundantly available at the cell surface than in other species. Therefore, to cause LAL gelation fewer cells of *N. gonorrhoeae* were required than of other gram-negative bacteria which might be encountered in urethral or cervical secretions (Spagna et al 1980). Thus, it appeared possible to simply dilute a specimen to a degree that contaminants or nongonococcal pathogens would no longer contribute enough endotoxin to cause a positive test, and still detect the relatively larger quantity of endotoxin produced by gonococci.

This principle has also been applied successfully by Spagna and colleagues (Table 4) in the examination of female cervical secretions in two separate studies (1980, 1981). In their first study, 94% of women with gonorrhea were detected by LAL testing and all patients without gonorrhea were found to

TABLE 4 *Summary of eleven studies using the LAL test for diagnosis of gonorrhea*

Specimens tested by LAL	Specimen dilution	Positive LAL tests		Reference
		Patients with gonorrhea*	Patients without gonorrhea*	
Male urethral exudates	1:200	100% (73/73)	4% (1/27)	Spagna et al (1979)
Cervical secretions	1:800	94% (17/18)	0 (0/22)	Spagna et al (1980)
Male urethral exudates	1:1600	100% (61/61)	4% (2/55)	Prior and Spagna (1981a)
Cervical secretions	1:800	100% (3/3)	– (0/0)	Spagna and Prior (1981)
Cervical secretions	1:50	71% (17/24)	7% (3/41)	Young et al (1981)
Male urethral exudates	1:400	99% (363/366)	3% (6/184)	Prior and Spagna (1981b)
Male urethral exudates	1:400-600	95% (163/172)	11% (3/28)	Chapel et al (1981)
Male urethral exudates	1:400-600	100% (56/56)	0 (0/15)	Nachum and Christiansen (1982)
Cervical secretions	1:2000	100% (4/4)	43% (19/44)	Hainer et al (1982)
Cervical secretions	Not defined	100% (71/71)	52% (23/44)	Prior and Spagna (1982)
Male urethral exudates	1:400	98% (66/67)	7% (4/58)	Prior and Spagna (1983)

* Presence or absence of *N. gonorrhoeae* documented by cultures

be negative. However, a similar study performed by Young et al (1981) did not provide such convincing data. Only 70% of women with gonorrhea yielded a positive LAL test at a relatively low dilution of 1:50, while 7% of women without any history of gonorrhea also yielded positive LAL reactions. However, the latter authors used a different lysate preparation than the one Prior and Spagna found most suitable (1981a). Non-organic-solvent extracted lysate was found to be more sensitive to the presence of native gonococcal endotoxin (Prior and Spagna 1981a) than the extracted lysate as used by Young et al (1981).

Conversely, Hainer et al (1982) found that the LAL test was not sufficiently specific to be used in a low-incidence, largely asymptomatic female population. While sensitivity for recognition of women with gonorrhea was not questioned (4/4 were positive), use of a 1:2000 dilution of cervical secretions did not prevent the alarming incidence of 43% false-positive LAL tests in culture-negative women. A subsequent communication from Prior and Spagna (1982) in fact reported a similar rate of false-positivity (52%) in cervical secretions from women who had had sexual contact with men who had culture-proven gonorrhea. However, Prior and Spagna (1982) chose to take the opposite view that a 48% negative LAL test rate among culture-negative women excluded them from further epidemiologic investigation or inappropriate antibiotic therapy.

It appeared that the LAL test could more easily be employed for the examination of male urethral exudates than for use on endocervical secretions of females. Prior and Spagna (1981b) described the use of a 'kit' for examining male urethral fluids in a group of 550 men. Their test device included a syringe for sample collection, a dilution reservoir to provide a 1:400 dilution, and a single test vial of LAL. By the use of this procedure and a 30-minute LAL incubation period, sensitivity of 99.2% and specificity of 96.4% was achieved. Concomitant gram-stain examination of the same fluids provided lower sensitivity (96.4%), but greater specificity (99.5%). However, there was not a statistically significant difference in the abilities of the LAL test and gram-stain to predict eventual gonococcal culture results ($p > 0.05$). The use of this test device was said to provide the opportunity for a private physician to make a presumptive diagnosis of gonorrhea in his office within a 30-minute period without need of a microscope and staining supplies. However, a subsequent group of investigators which evaluated the same test kit provided somewhat less encouraging results. Chapel et al (1981) detected 95% of male urethral samples which contained gonococci, although

11% of culture-negative samples also produced positive LAL tests. In fact, the gram-stained smears of the same specimens provided both greater sensitivity and specificity than the LAL test, which led the authors to conclude that the LAL test did not appear useful for the presumptive diagnosis of gonorrhea in men. Nachum and Christiansen (1982), however, found the LAL test kit to have a very high predictive value for detection of gonorrhea in a group of homosexual men seen in a clinic for diagnosis of sexually transmitted diseases. In this setting, neither false-positive nor false-negative tests were encountered.

Lastly, Prior and Spagna (1983) evaluated the use of a 10-minute chromogenic LAL test for examination of male urethral exudate samples. Use of the synthetic chromogenic lysate was shown to produce results similar to those obtained by gelation LAL testing and by gram-stain examination of the same specimens. However, the color endpoint obtained with the chromogenic LAL was said to be faster and easier to interpret and not prone to mechanical disturbances during incubation.

The diagnosis of gonorrhea by LAL is a new and potentially useful application of the test. The majority of studies and especially those of Prior and Spagna suggest that the LAL test could be considered as a more objective means of establishing a presumptive diagnosis of gonorrhea than the interpretation of a gram-stained smear. If compared directly with use of the gram-stain for this purpose, the sensitivity of the LAL test was generally impressive. However, the greater specificity of the gram-stain, especially in the hands of an experienced microscopist, could be extremely important in light of the social and medical-legal implications of a diagnosis of gonorrhea. The studies of Chapel et al (1981) on male urethral exudates and those of Hainer et al (1982) and Prior and Spagna (1982) cast serious doubt on the future success of the LAL test for the purpose, on the basis of specificity which is less than the current standard test (the gram-stain).

3.5. Detection of gram-negative bacteria in ocular infections

The possibility of applying the LAL assay to diagnosis of ocular diseases was first noted by Poirier and Jorgensen (1977) when they detected the endotoxin of *Pseudomonas aeruginosa* in the ulcer bed of a patient's infected cornea. A minute sample of corneal ulcer material obtained with a Kimura spatula was emulsified directly in a tube of LAL. The corneal sample caused abrupt gelation of the lysate within 30 minutes of collection. Subsequent to this, the use of the LAL test for recognition of gram-negative bacteria in corneal ulcers has proved to be another new and successful application (Wolters et al 1979).

It has also been found to be possible to test minute volumes of intraocular fluid for the presence of gram-negative bacteria by this method (Ellison 1978; McBeath et al 1978). The principal advantages of the LAL test for this purpose are the speed with which results become available (since gram-negative ocular infections may lead to irreversible damage within a few hours), and the objectivity of the LAL test, since stained smears obtained from ocular lesions or fluids may be especially difficult to interpret. In this regard, the LAL test may be considered only a rapid adjunctive technique to provide useful information before the results of cultures can be known. Suitable means for detecting gram-positive bacteria, fungi, and viruses must also be provided to establish the presence of microorganisms which do not produce endotoxin.

3.6. Other clinical applications of the LAL test

Newer applications of the LAL test continue to be reported. In virtually any infectious process in which it is important to distinguish between gram-positive and gram-negative bacteria, or between septic and nonseptic conditions, there is a potential use for the test. Berstein et al (1980) found the LAL test to be useful for detecting gram-negative bacteria in middle ear fluids of patients with otitis media. The prominence of *Haemophilus influenzae* and possibly *Branhamella catarrhalis* in the etiology of otitis media in children makes this another plausible application of the test.

It is more difficult to ascertain the potential utility of LAL testing of joint fluids. The etiology of pyogenic arthritis is itself difficult to establish since bacterial cultures are sometimes negative and smear results are very difficult to interpret. As in endotoxemia, the lack of a reliable comparative test makes it difficult to assess the use of LAL for this purpose. In both the series reported by Tuazon et al (1977), and that of Elin et al (1978), all patients with septic arthritis due to gram-negative bacteria had positive LAL tests of their joint fluids. It is more difficult to establish whether positive LAL tests on joint fluids from which gram-negative bacteria were not grown represent false-positive LAL tests, or instead, underscore the difficulty of confirming the diagnosis by conventional means. All patients with positive LAL assays in both studies (Elin et al 1978; Tuazon et al 1977) had increased numbers of leukocytes in their joint fluids and concomitantly depressed glucose levels – two findings which are strongly suggestive of bacterial infection, despite negative culture results. However, as in the use of the LAL test for detection of endotoxemia, a lack of correlation with culture results compromises the usefulness of the LAL for this purpose.

Another area of application of the LAL test is that of hospital epidemiology. The concept of LAL testing of parenteral fluids for pyrogens may be readily applied to in-use contamination of fluids in a hospital setting. Jorgensen and Smith (1973b) used the assay to rapidly detect an intravenous fluid preparation that had been inadvertently contaminated with *Pseudomonas aeruginosa* during its admixture in the hospital. Elin et al (1975a) made similar use of the test to recognize a contaminated bottle of sodium chloride solution used for washing frozen red blood cells. LAL testing of solutions and source waters used in hemodialysis has also been extremely useful for detecting sources of gram-negative bacterial contamination (Carson et al 1979; Hindman et al 1975). Rusmin and DeLuca (1975) demonstrated periodic release of endotoxin from bacteria retained on intravenous in-line filters when antibiotics were flushed through the filter. Several other investigators have likewise used LAL testing to explain pyrogenic reactions due to fluids or devices which have been rendered sterile, but not pyrogen-free (Fumarola 1977; Kundsin and Walter 1980; Siegel et al 1976).

4. SUMMARY

The conclusions which may be drawn from a review of the clinical applications of the LAL test are that a number of potentially successful uses have been described since the test was introduced in the late 1960s. They have all benefited from the speed, sensitivity, and general specificity of the LAL test for bacterial endotoxin, whether it be bound endotoxin from viable gram-negative bacteria or free endotoxin as the residual evidence of pre-existent or periodic bacterial growth. In all of these applications, the LAL test has shown some distinct improvement over conventional methods of diagnosis. However, none of these applications have replaced conventional methods with strict reliance on LAL testing. Instead, the LAL test is a powerful adjunctive means of discovering the presence of gram-negative bacteria in a variety of clinical circumstances. It has several current valid diagnostic uses (meningitis, ocular infections, bacteriuria) and may eventually lead to a better understanding of other pathophysiologic consequences of endotoxins, that is, endotoxemia.

REFERENCES

Baek L (1983) New, sensitive rocket immunoelectrophoretic assay for measurement of the reaction between endotoxin and Limulus amoebocyte lysate. *J. Clin. Microbiol. 17*, 1013-1020.

Bang FB (1958) A bacterial disease of *Limulus polyphemus. Bull. Johns Hopkins Hosp. 98*, 325-351.

Berman NS, Siegel SE, Nachum R, Lipsey A, Leedom J (1976) Cerebrospinal fluid endotoxin concentrations in gram-negative bacterial meningitis. *J. Pediatr. 88*, 553-556.

Berstein JM, Praino MD, Neter E (1980) Detection of endotoxin in ear specimens from patients with chronic otitis media by means of the Limulus amebocyte lysate test. *Can. J. Microbiol. 26*, 546-548.

Brody L, Webster MC, Kark RM (1968) Identification of elements of urinary sediment with phase contrast microscopy. *J. Am. Med. Assoc. 206*, 1777-1781.

Butler T, Levin J, Cu DQ, Walker RI (1973) Bubonic plague: detection of endotoxemia with the Limulus test. *Ann. Intern. Med. 79*, 642-646.

Butler T, Levin J, Linh NN, Chau DM, Adickman M, Arnold K (1976) *Yersinia pestis* infection in Vietnam. II. Quantitative blood cultures and detection of endotoxin in the cerebrospinal fluid of patients with meningitis. *J. Infect. Dis. 133*, 493-499.

Butler T, Bell WR, Levin J, Linh NN, Arnold K (1978) Typhoid fever. Studies of blood coagulation, bacteremia, and endotoxemia. *Arch. Intern. Med. 138*, 407-410.

Caridis DT, Reinhold RB, Woodruff PWH, Fine J (1972) Endotoxaemia in man. *Lancet 2*, 1381-1385.

Carson LA, Petersen NJ, Favero MS (1979) Use of the Limulus amoebocyte lysate assay system for detection of bacterial endotoxin in fluids associated with hemodialysis procedures. In: Cohen E (Ed), *Biomedical Applications of the Horseshoe Crab (Limuliedae)*, pp 453-464. Alan R. Liss, New York.

Chapel TA, Adcock M, Smith B, Barnes T, Goodman M (1981) Evaluation of the limulus amebocyte lysate assay for presumptive diagnosis of gonorrhea in men at a clinic for sexually transmitted diseases. *Sex. Transm. Dis. 8*, 175-178.

Cooper JF, Neely ME (1980) Validation of the LAL test for end-product evaluation. *Pharm. Technol. 4*, 72-79.

Cooperstock MS, Tucker RP, Baublis JV (1975) Possible pathogenic role of endotoxin in Reye's syndrome. *Lancet 2*, 1272-1274.

Cuevas P, Fine J (1972) Role of intraintestinal endotoxin in death from peritonitis. *Surg. Gynecol. Obstet. 134*, 953-957.

Das J, Schwartz AA, Folkman J (1973) Clearance of endotoxin by platelets: role in increasing the accuracy of the Limulus gelation test and in combating experimental endotoxemia. *Surgery 74*, 235-240.

DuBose D, Lemaire M, Brown J, Wolfe D, Hamlet M (1978) Survey for positive *Limulus* amoebocyte lysate test in plasma from humans and common research animals. *J. Clin. Microbiol. 7*, 139-141.

DuBose DA, Lemaire M, Basamania K, Rowlands J (1980) Comparison of plasma extraction techniques in preparation of samples for endotoxin testing by the Limulus amoebocyte lysate test. *J. Clin. Microbiol. 11*, 68-72.

Dumont JN, Anderson E, Winner G (1966) Some cytologic characteristics of the hemocytes of Limulus during clotting. *J. Morphol. 119*, 181-208.

Dyson D, Cassady G (1976) Use of Limulus lysate for detecting gram-negative neonatal meningitis. *Pediatrics 58*, 105-109.

Elin RJ, Wolff SM (1973) Nonspecificity of the limulus amebocyte lysate test: positive reactions with polynucleotides and proteins. *J. Infect. Dis. 128*, 349-352.

Elin RJ, Lundberg WB, Schmidt PJ (1975a) Evaluation of bacterial contamination in blood processing. *Transfusion 15*, 260-265.

Elin RJ, Robinson RA, Levine AS, Wolff SM (1975b) Lack of clinical usefulness of the Limulus test in the diagnosis of endotoxemia. *N. Engl. J. Med. 293*, 521-524.

Elin RJ, Knowles R, Barth VF, Wolff SM (1978) Lack of specificity of the Limulus lysate test in the diagnosis of pyogenic arthritis. *J. Infect. Dis. 137*, 507-513.

Ellison AC (1978) The Limulus lysate test. *Arch. Ophthalmol. 96*, 1268-1271.

Feldman S, Pearson TA (1974) The Limulus test and gram-negative bacillary sepsis. *Am. J. Dis. Child. 128*, 172-174.

Feldman WE (1976) Concentrations of bacteria in cerebrospinal fluid of patients with bacterial meningitis. *J. Pediatr. 88*, 549-552.

Fink PC, Lehr L, Urbaschek RM, Kozak J (1981) Limulus amebocyte lysate test for endotoxemia: investigations with a femtogram sensitive spectrophotometric assay. *Klin. Wochenschr. 59*, 213-218.

Fleishman J, Fowlkes F (1982) A comparison of pyrogenicity of bacterial endotoxins from a variety of Gram-negative bacteria as determined by the LAL test. *Prog. Clin. Biol. Res. 93*, 131-140.

Fossard DP, Kakkar VV, Elsey PA (1974) Assessment of Limulus test for detecting endotoxaemia. *Br. Med. J. 2*, 465-468.

Fumarola D (1977) Possible endotoxin contamination in some antitumor agent preparations: study with the Limulus amebocyte lysate test. *Farmaco (Pavia) 32*, 444-448.

Fumarola D, Munno I, Monno R, Miragliotta G (1980) Lipopolysaccharides from *Rickettsiaceae*: Limulus endotoxin assay and pathogenic mediators in rickettsiosis. *Acta Virol. 24*, 155.

Gaffin SL, Obedeanu N, Merzbach D (1979) The Limulus amebocyte lysate (LAL) test for endotoxin in human plasma. *Thromb. Haemostasis 42*, 808-809.

Galloway RE, Levin J, Butler T, Naff GB, Goldsmith GH, Saito H, Awoke S, Wallace CK (1977) Activation of protein mediators of inflammation and evidence for endotoxemia in *Borrelia recurrentis* infection. *Am. J. Med. 63*, 933-938.

Garibaldi RA, Allman GW, Larsen DH, Smith CB, Burke JP (1973) Detection of endotoxemia by the Limulus test in patients with indwelling urinary catheters. *J. Infect. Dis. 128*, 551-554.

Gelfand JA, Elin RJ, Berry FW, Frank MM (1976) Endotoxemia associated with the Jarisch-Herxheimer reaction. *N. Engl. J. Med. 295*, 211-213.

Goldstein JA, Reller LB, Wang W-LL (1976) Limulus amoebocyte lysate test in neonates. *Am. J. Clin. Pathol. 66*, 1012-1015.

Hainer BL, Danylchuk P, Cooper J, Weart CW (1982) Limulus lysate assay in detection of gonorrhea in women from a low-incidence population. *Am. J. Obstet. Gynecol. 144*, 67-71.

Highsmith AK, Mackel DC, Baine WB, Anderson RL, Fraser DW (1978) Observations of endotoxin-like activity associated with the Legionnaires' disease bacterium. *Curr. Microbiol. 1*, 315-317.

Hindman SH, Favero MS, Carson LA, Petersen NJ, Schonberger LB, Solano JT (1975) Pyrogenic reactions during haemodialysis caused by extramural endotoxin. *Lancet 2*, 732-734.

Hollander VP, Harding WC (1976) A sensitive spectrophotometric method for measurement of plasma endotoxin. *Biochem. Med. 15*, 28-33.

Holme R, Solum NO (1973) Electron microscopy of the gel protein formed by clotting of *Limulus polyphemus* hemocyte extracts. *J. Ultrastruct. Res. 44*, 329-338.

Iwanaga S, Morita T. Harada T, Nakamura S, Niwa M, Takada K, Kimura T, Sakakibara S (1978) Chromogenic substrate for horseshoe crab clotting enzyme. Its application for the assay of bacterial endotoxins. *Haemostasis 7*, 183-188.

Jacob AI, Goldberg PK, Bloom N, Degenshein GA, Kozinn PJ (1977) Endotoxin and bacteria in portal blood. *Gastroenterology 72*, 1268-1270.

Johnson KJ, Ward PA, Goralnick S, Osborn MJ (1977) Isolation from human serum of an inactivator of bacterial lipopolysaccharide. *Am. J. Pathol. 88*, 559-574.

Jones RJ, Roe EA (1979) Measurement of endotoxins with the Limulus test in burned patients. *J. Hyg. 83*, 151-156.

Jorgensen JH, Alexander GA (1981) Automation of the Limulus amoebocyte lysate test by using the Abbott MS-2 microbiology system. *Appl. Environ. Microbiol. 41*, 1316-1320.

Jorgensen JH, Alexander GA (1982) Rapid detection of significant bacteriuria by use of an automated *Limulus* amoebocyte lysate assay. *J. Clin. Microbiol. 16*, 587-589.

Jorgensen JH, Jones PM (1975) Comparative evaluation of the Limulus assay and the direct gram stain for detection of significant bacteriuria. *Am. J. Clin. Pathol.*

63, 142-148.

Jorgensen JH, Lee JC (1978) Rapid diagnosis of gram-negative bacterial meningitis by the Limulus endotoxin assay. *J. Clin. Microbiol. 7*, 12-17.

Jorgensen JH, Smith RF (1973a) Preparation, sensitivity, and specificity of Limulus lysate for endotoxin assay. *Appl. Microbiol. 26*, 43-48.

Jorgensen JH, Smith RF (1973b) Rapid detection of contaminated intravenous fluids using the Limulus in vitro endotoxin assay. *Appl. Microbiol. 26*, 521-524.

Jorgensen JH, Smith RF (1974) Measurement of bound and free endotoxin by the Limulus assay. *Proc. Soc. Exp. Biol. Med. 146*, 1024-1031.

Jorgensen JH, Carvajal HF, Chipps BE, Smith RF (1973) Rapid detection of gram-negative bacteriuria by use of the Limulus endotoxin assay. *Appl. Microbiol. 26*, 38-42.

Kundsin RB, Walter CW (1980) Detection of endotoxin on sterile catheters used for cardiac catheterization. *J. Clin. Microbiol. 11*, 209-212.

Levin J, Bang FB (1964) A description of cellular coagulation in Limulus. *Bull. Johns Hopkins Hosp. 115*, 337-345.

Levin J, Bang FB (1968) Clottable protein in Limulus: its localization and kinetics of its coagulation by endotoxin. *Thromb. Diath. Haemorrh. 19*, 186-197.

Levin J, Poore TE, Zauber NP, Oser RS (1970a) Detection of endotoxin in the blood of patients with sepsis due to gram-negative bacteria. *N. Engl. J. Med. 283*, 1313-1316.

Levin J, Tomasulo PA, Oser RS (1970b) Detection of endotoxin in human blood and demonstration of an inhibitor. *J. Lab. Clin. Med. 75*, 903-911.

Levin J, Poore TE, Young NS, Margolis S, Zauber NP, Townes AS, Bell WR (1972) Gram-negative sepsis: detection of endotoxemia with the Limulus test. With studies of associated changes in blood coagulation, serum lipids, and complement. *Ann. Int. Med. 76*, 1-7.

Liang S-M, Liu T-Y (1982) Studies on the Limulus coagulation system: inhibition of activation of the proclotting enzyme by dimethyl sulfoxide. *Biochem. Biophys. Res. Commun. 105*, 553-559.

Mahalanabis D, Jalan KN, Maitra TK, Agarwal SK, Chakraborty ML (1979) In search of a sister species of *Limulus polyphemus* in the Bay of Bengal: clottable protein in the amoebocyte lysate of the *Carcinoscorpius rotunda cauda*. *Indian J. Med. Res. 70*, 36-40.

Martinez-G LA, Quintiliani R, Tilton RC (1973) Clinical experience on the detection of endotoxemia with the Limulus test. *J. Infect. Dis. 127*, 102-105.

McBeath J, Forster RK, Rebell G (1978) Diagnostic limulus lysate assay for endophthalmitis and keratitis. *Arch. Ophthalmol. 96*, 1265-1267.

McCracken GH, Sarff LD (1976) Endotoxin in cerebrospinal fluid. Detection in neonates with bacterial meningitis. *J. Am. Med. Assoc. 235*, 617-620.

McCullough KZ, Scolnick SA (1976) Effect of semisynthetic penicillins on the

Limulus lysate test. *Antimicrob. Agents Chemother. 9*, 856-858.

McLeod C, Katz W (1981) A rapid method for the detection of Gram-negative bacterial endotoxins in whole blood. *J. Biol. Stand. 9*, 299-306.

Mikami T, Nagase T, Matsumoto T, Suzuki S, Suzuki M (1982) Gelation of Limulus amoebocyte lysate by simple polysaccharides. *Microbiol. Immunol. 26*, 403-409.

Murer EH, Levin J, Holme R (1975) Isolation and studies of the granules of the amebocytes of *Limulus polyphemus*, the horseshoe crab. *J. Cell Physiol. 86*, 533-542.

Nachum R, Christiansen RO (1982) Rapid presumptive diagnosis of gonococcal urethritis in males: evaluation of a prototype Limulus lysate test kit. *Med. Lab. Sci. 39*, 83-85.

Nachum R, Shanbrom E (1981) Rapid detection of gram-negative bacteriuria by Limulus amoebocyte lysate assay. *J. Clin. Microbiol. 13*, 158-162.

Nachum R, Lipsey A, Siegel SE (1973) Rapid detection of gram-negative bacterial meningitis by the Limulus lysate test. *N. Engl. J. Med. 289*, 931-934.

Nakamura S, Iwanaga S, Harada T, Niwa M (1976) A clottable protein (coagulogen) from amoebocyte lysate of Japanese horseshoe crab (*Tachypleus tridentatus*). Its isolation and biochemical properties. *J. Biochem. 80*, 1011-1021.

Nakamura S, Morita T, Iwanaga S, Niwa M, Takahashi K (1977) A sensitive substrate for the clotting enzyme in horseshoe crab hemocytes. *J. Biochem. 81*, 1567-1569.

Nandan R, Nakashima CY, Brown DR (1977) Detection of endotoxins in human blood and plasma. An improved in-vitro pyrogen test. *Clin. Chem. 23*, 2080-2084.

Neter E (1965) Evaluation of the tetrazolium test for the diagnosis of significant bacteriuria. *J. Am. Med. Assoc. 192*, 769.

Oberle MW, Graham GG, Levin J (1974) Detection of endotoxemia with the Limulus test: preliminary studies in severely malnourished children. *J. Pediatr. 85*, 570-573.

Pearson FC, Weary M (1980) The significance of limulus amebocyte lysate test specificity on the pyrogen evaluation of parenteral drugs. *J. Parenteral Drug Assoc. 34*, 103-108.

Pezzlo MT, Tan GL, Peterson EM, De la Maza LM (1982) Screening of urine cultures by three automated systems. *J. Clin. Microbiol. 15*, 468-474.

Platica M, Harding W, Hollander VP (1978) Dithiols stimulate endotoxin in the Limulus reaction. *Experientia 34*, 1154-1155.

Poirier R, Jorgensen JH (1977) Endotoxin assay for rapid diagnosis of Pseudomonas corneal ulcer. *Lancet 2*, 85-86.

Prior RB, Spagna VA (1981a) Response of several *Limulus* lysates to native endotoxin present in gonococcal and non-gonococcal urethral exudates from human males. *J. Clin. Microbiol. 13*, 167-170.

Prior RB, Spagna VA (1981b) Application of a limulus test device in rapid evaluation of gonococcal and nongonococcal urethritis in males. *J. Clin. Microbiol. 14*, 256-260.

Prior RB, Spagna VA (1982) Rapid evaluation of female patients exposed to gonorrhea by use of the Limulus lysate test. *J. Clin. Microbiol. 16*, 487-489.

Prior RB, Spagna VA (1983) Rapid evaluation of gonococcal and nongonococcal urethritis in men with *Limulus* amoebocyte lysate and a chromogenic substrate. *J. Clin. Microbiol. 17*, 485-488.

Ravin HA, Rowley D, Jenkins C, Fine J (1960) On the absorption of bacterial endotoxin from the gastrointestinal tract of the normal and shocked animal. *J. Exp.Med. 112*, 783-792.

Reinhold RB, Fine J (1971) A technique for quantitative measurement of endotoxin in human plasma. *Proc. Soc. Exp. Biol. Med. 137*, 334-340.

Robinson JA, Klodnycky ML, Loeb HS, Racic MR, Gunnar RM (1975) Endotoxin, prekallikrein, complement and systemic vascular resistance. Sequential measurements in man. *Am. J. Med. 59*, 61-67.

Rojas-Corona RR, Skarnes R, Tamakuma S, Fine J (1969) The Limulus coagulation test for endotoxin. A comparison with other assay methods. *Proc. Soc. Exp. Biol. Med. 132*, 599-601.

Ross S, Rodriguez W, Controni G, Korengold G, Watson S, Khan W (1975) Limulus lysate test for gram-negative bacterial meningitis. *J. Am. Med. Assoc. 233*, 1366-1369.

Rowe MI, Buckner DM, Newmark S (1975) The early diagnosis of gram-negative septicemia in the pediatric surgical patient. *Ann. Surg. 182*, 280-286.

Rusmin S, DeLuca PP (1975) Effect of antibiotics and osmotic change on the release of endotoxin by bacteria retained on intravenous inline filters. *Am. J. Hosp. Pharm. 32*, 378-380.

Scheifele DW, Melton P, Whitchelo V (1981) Evaluation of the Limulus test for endotoxemia in neonates with suspected sepsis. *J. Pediatr. 98*, 899-903.

Schleef RR, Kenney DM, Shepro D (1979) Effect of sodium chloride on Limulus amoebocyte lysate. Inhibition of endotoxin activation of procoagulase. *Thromb. Haemostasis 41*, 329-336.

Schleef RR, Shepro D, Kenney DM (1982) Effect of urea on the endotoxin activated clotting system of *Limulus polyphemus. Thromb. Res. 27*, 729-735.

Schweinburg FB, Fine J (1960) Evidence for a lethal endotoxemia as the fundamental features of irreversibility in three types of traumatic shock. *J. Exp. Med. 112*, 793-800.

Sibbald WJ (1976) Bacteremia and endoxemia: a discussion of their roles in the pathophysiology of gram-negative sepsis. *Heart Lung 5*, 765-771.

Siegel SE, Nachum R, Leimbrock S, Karon M (1976) Detection of bacterial endotoxin in antitumor agents. *Cancer Treat. Rep. 60*, 9-15.

Spagna VA, Prior RB (1981) Application of the Limulus lysate assay in evaluation of disseminated gonorrhea in women. *Sex. Trans. Dis. 8*, 18-20.

Spagna VA, Prior RB, Perkins RL (1979) Rapid presumptive diagnosis of gonococcal urethritis in men by the Limulus lysate test. *Br. J. Vener. Dis. 55*, 179-182.

Spagna VA, Prior RB. Perkins RL (1980) Rapid presumptive diagnosis of gonococcal cervicitis by the Limulus lysate assay. *Am. J. Obstet. Gynecol. 137*, 595-599.

Steindler KA, Tsuji K, Enzinger RM (1981) Potentiating effect of calcium gluconate on the Limulus amebocyte lysate (LAL) gelation-endpoint assay for endotoxin. *J. Parenteral Sci. Tech. 35*, 242-247.

Stumacher RJ, Kovnat MJ, McCabe WR (1973) Limitations on the usefulness of the Limulus assay for endotoxin. *N. Engl. J. Med. 288*, 1261-1264.

Sullivan JD, Watson SW (1975) Inhibitory effect of heparin on the Limulus test for endotoxin. *J. Clin. Microbiol. 2*, 151.

Suzuki M, Mikami T, Matsumoto T, Suzuki S (1977) Gelation of Limulus lysate by synthetic dextran derivatives. *Microbiol. Immunol. 21*, 419-425.

Tai JY, Liu T-Y (1977) Studies on Limulus amoebocyte lysate. Isolation of pro-clotting enzyme. *J. Biol. Chem. 252*, 2178-2181.

Tai JY, Seid RC, Huhn RD, Liu T-Y (1977) Studies on Limulus amoebocyte lysate II. *J. Biol. Chem. 252*, 4773-4776.

Takagi K, Morita A, Tamura H, Nakahara C, Tanaka S, Fujita Y, Kawai T (1981) Quantitative measurement of endotoxin in human blood using synthetic chromogenic substrate for horseshoe crab clotting enzyme: a comparison of methods of blood sampling and treatment. *Thromb. Res. 23*, 51-57.

Tamura H, Obayashi T, Takagi K, Tanaka S, Nakahara C, Kawai T (1982) Perchloric acid treatment of human blood for quantitative endotoxin assay using synthetic chromogenic substrate for horseshoe crab clotting enzyme. *Thromb. Res. 27*, 51-57.

Tarao K, So K, Moroi T, Ikeuchi T, Suyama T, Endo O, Fukushima K (1977) Detection of endotoxin in plasma and ascitic fluid of patients with cirrhosis: its clinical significance. *Gastroenterology 73*, 539-542.

Teller JD, Kelly KM (1979) A turbidimetric Limulus amebocyte assay for the quantitative determination of gram-negative bacterial endotoxin. In: Cohen E (Ed.), *Biomedical Applications of the Horseshoe Crab (Limulidae)*, pp 423-434. Alan R. Liss, New York.

The United States Pharmacopeia, 20th Rev. Ed. (1980) p 888. Mack Publishing Co., Easton PA.

Tuazon CA, Perez AA, Elin RJ, Sheagren JN (1977) Detection of endotoxin in cerebrospinal and joint fluids by Limulus assay. *Arch. Intern. Med. 137*, 55-56.

Usawattanakul W, Tharavanij S, Limsuwan A (1979) Tachypleus lysate test for endotoxin in patients with Gram-negative bacterial infections. *Southeast Asian J. Trop. Med. Public Health 10*, 13-17.

Walker RI (1978) The contribution of intestinal endotoxin to mortality in hosts with compromised resistance: a review. *Exp. Hematol. 6*, 172-184.

Weary ME, Donohue G, Pearson FC, Story K (1980) Relative potencies of four reference endotoxin standards as measured by the Limulus amoebocyte lysate and USP rabbit pyrogen tests. *Appl. Environ. Microbiol. 40*, 1148-1151.

Webster CJ (1980) Principles of a quantitative assay for bacterial endotoxins in blood that uses Limulus lysate and a chromogenic substrate. *J. Clin. Microbiol. 12*, 644-650.

Wilkinson SP, Arroyo V, Gazzard BG, Moodie H, Williams R (1974) Relation of renal impairment and haemorrhagic diathesis to endotoxaemia in fulminant hepatic failure. *Lancet 1*, 521-524.

Wolters RW, Jorgensen JH, Calzada E, Poirier RH (1979) Limulus lysate assay for early detection of certain gram-negative corneal infections. *Arch. Ophthalmol. 97*, 875-877.

Yin ET (1975) Endotoxin, thrombin, and the Limulus amebocyte lysate test. *J. Lab. Clin. Med. 86*, 430-434.

Young LS (1975) Opsonizing antibodies, host factors, and the Limulus assay for endotoxin. *Infect. Immun., 12*, 88-92.

Young NS, Levin J, Prendergast NA (1972) An invertebrate coagulation system activated by endotoxin: evidence for enzymatic mediation. *J. Clin. Invest. 51*, 1790-1797.

Young H, Sarafian SK, McMillan A (1981) Reactivity of the Limulus lysate assay with uterine cervical secretions. A preliminary evaluation. *Br. J. Vener. Dis. 57*, 200-203.

Handbook of Endotoxin, Vol. 4: Clinical Aspects of Endotoxin Shock
R.A. Proctor, editor
© Elsevier Science Publishers B.V., 1986

CHAPTER 5

Role of antibody in the prevention and pathogenesis of endotoxin and gram-negative septic shock

RICHARD A. PROCTOR

1. INTRODUCTION

Morbidity and mortality from gram-negative rod bacteremia and endotoxemia has increased in frequency in hospitalized patients over the last 40 years (reviewed by Young 1979). Several organisms classically associated with overwhelming gram-negative sepsis, for instance *Yersinia pestis*, *Salmonella typhi*, and *Neisseria meningitidis*, have been replaced by *Klebsiella*, *Enterobacter*, and *Pseudomonas* species as the agents causing shock and death in hospitalized patients. Before the advent of sulfonamides, *Escherichia coli* and *Salmonella* species were the only gram-negative organisms seen with any frequency at Boston City Hospital (McGowan et al 1975). However, between 1957 and 1972, the incidence of gram-negative bacteremia increased dramatically and organisms such as *Klebsiella pneumoniae* and *Enterobacter* began to be seen frequently outside the setting of Friedlander's pneumonia (McGowan et al 1975). Similar changes were seen at other hospitals (McCabe and Jackson, 1962a, 1962b; Rogers 1959; Young 1979).

A number of factors have caused this rise in hospital-acquired gram-negative infections. Antimicrobial agents have accounted for this increase in at least two ways. First, gram-negative bacilli superinfection may follow the use of antimicrobial agents (Waisbern 1951). Second, patients who would have succumbed to infections caused by gram-positive cocci, now become infected with gram-negative bacilli; for example, without antibiotics, patients with major burns often would have died of *Streptococcus pyogenes* sepsis (Burke 1961) but now become infected with gram-negative rods (Glew et al 1976). Coinciding with the advent of antibiotics has been a host of medical and surgical advances which also predispose a patient to nosocomial gram-negative bacillary infection. Advances such as respirators, hemodialysis, transplantation, cancer chemotherapy, burn units, intensive care units, greater use of invasive diagnostic techniques, intravascular monitoring devices, and many other advances have all increased both survival and the frequency of nosocomial gram-negative infections. Thus, many of the gram-negative bacillary infections are diseases of medical progress.

Despite the advances in antibiotics, fluid, and pressor therapy, the mortality rate has remained distressingly high in gram-negative rod infection, especially in patients who have had a rapidly fatal underlying disease (Bryant et al 1971; Christy 1971; Myerowitz et al 1971). Because of the high mortality and the continued therapeutic failures, efforts have been directed toward immunotherapy of gram-negative sepsis and endotoxic shock.

2. HISTORICAL PERSPECTIVE

Effective immunotherapy for gram-negative rod infections dates from the earliest beginnings of immunology. Pfeiffer and Kolle (1896) transferred serum from patients convalescing with typhoid fever to guinea pigs. This serum protected the guinea pigs from a lethal challenge with *S. typhi* whereas control serum did not. In 1938, Boivin and Mesrobeanu produced antibodies in rabbits to *Shigella* endotoxins. The immune sera given subcutaneously to mice did not protect them from subsequent *Shigella* challenge. Weil et al (1939) found similar results with typhoid. By combining *S. typhi* endotoxin with an excess of antibody, Morgan (1941) was able to inhibit the local Shwartzman reaction. Creech et al (1948) reported that intraperitoneal injection of rabbit antisera protected mice against lethal intraperitoneal doses of *Serratia marcescens* endotoxin. Abernathy and Spink (1958) were the first to show that human convalescent serum (*Brucella melitensis* infection) was able to neutralize endotoxin. Freedman (1959) demonstrated that serum, even from heterologous bacterial strains, provided passive protection against the febrile response to endotoxin in rabbits. He was unable to relate this to the concentration of antibody against the endotoxins used, but in 1963, Watson and Kim showed that antibodies to endotoxin did inhibit the febrile response. Antibodies were shown to be effective against heterologous endotoxins (McCabe and Greely 1973; McCabe et al 1973b).

These early data suggesting heterologous protection allowed hope for the development of a vaccine which might recognize a common element among the wide diversity of antigens expressed by gram-negative bacilli. If just *E. coli* and *K. pneumoniae* are considered, over 200 serotypes exist (Edwards and Ewing 1972). Although 10 serotypes of *E. coli* cause about 60% of infections (Ørskov and Ørskov 1975), a vaccine against these serotypes would still leave the host unprotected for over 100 serotypes which cause the other 40% of *E. coli* infections. In addition, production of a vaccine which would include even a bare majority of serotypes of all the Enterobacteriaceae would prove a monumental task. A common antigen among the Enterobacteriaceae was described by Kunin in 1963 (reviewed by Mäkelä and Mayer 1976). However, antibody to the 'common antigen' did not correlate with decreased shock or increased survival in patients septic with gram-negative rods (Mc-Cabe et al 1973a, 1973b).

Another avenue for vaccine production followed from two lines of work. In studying the structure of endotoxin, a common 'core' region was found which demonstrated a high degree of similarity between diverse groups of organisms (reviewed by Heath et al 1966; Lüderitz et al 1966; Osborn 1966). At about the same time, 'rough' mutants of *Salmonella* were first isolated (Beckmann et al 1964; Subbaiah and Stocker 1964). These mutants were designated 'rough' because colonies on solid media were ragged and uneven rather than smooth and mucoid. The connection between these rough mutants and the core structure of endotoxin is shown in Figure 1 (see Chapters 4–10, Volume 1 of this series for more details of endotoxin structure). The

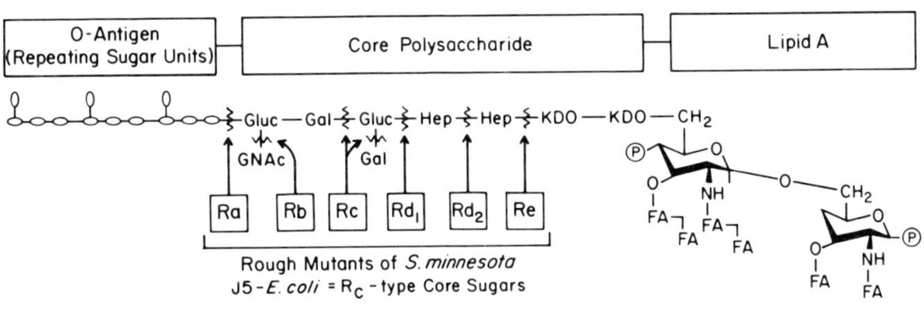

Fig. 1 *Rough mutants and the structure of endotoxin.*

outer membrane of gram-negative bacilli contains endotoxin, which can be divided into three major regions: (a) a lipid portion, lipid A, extending into the outer lipid membrane; (b) a 'core polysaccharide' region, which contains similar elements in a wide variety of species; and (c) 'O' polysaccharides. The 'O' antigens are repeating oligosaccharide units containing 3 or 4 different hexoses. These O antigens impart serotypic identity and 'smoothness' to the colonies. Colonies become progressively rougher, with increasing loss of the oligosaccharides. The O antigens are lost through mutations that deprive the bacteria of enzymes needed to synthesize the hexoses or insert them onto the core. A loss in the O antigens exposes the core polysaccharides because the O polysaccharides are the outermost portion of the organism. Mutants of *Salmonella minnesota* are designated 'Ra' for those organisms lacking O polysaccharides and 'Rb to Re' for even rougher organisms whose further mutations have led to a progressive loss of core polysaccharides. Thus the development of 'rough' organisms, along with the discovery of structures common to many gram-negative bacilli (lipid A and core polysaccharides), opened the possibility of producing a vaccine strain which would provide broad coverage.

3. PROTECTIVE MECHANISMS MEDIATED BY ENDOTOXIN-SPECIFIC ANTIBODY

Before the data are presented for the protective actions of anti-endotoxin antibody, mechanisms of antibody-mediated protection will be considered. The following discussion will consider antibodies directed against the O polysaccharide, the core region, and lipid A.

3.1. Neutralization and/or clearance of endotoxin

The most direct protective action of anti-endotoxin antibody would be neutralization and/or enhanced clearance of endotoxin from the host. Although anti-endotoxin antibodies, such as anti-O and anti-core antibodies, enhance clearance of homologous endotoxins (Tate et al 1966), heterologous endotoxins are not cleared at an enhanced rate (Braude and Douglas 1972). Nevertheless, the Shwartzman reaction and lethal effects of heterologous endotoxins were inhibited by anti-core antibodies produced against a rough *E. coli* (Braude and Douglas 1972). This indicates that anti-endotoxin antibodies can 'move aside' the O antigens to reach the core structures. O antigens are known to slow the clearance of endotoxins and smooth organisms in non-immunized animals (Chedid et al 1968) and to block even the action of specific O antisera by being free on the cell surface but unattached to lipid A (Goldman et al 1982). Antibodies raised against rough mutants, for example, the Rc mutant of *E. coli* (J5 mutant of *E. coli* O111:B4) or the Re mutant of *S. minnesota* R595, can be absorbed by smooth organisms which are of homologous or heterologous serotypes and can interact with endotoxins derived from smooth organisms (Chedid et al 1968; Ito et al 1980; Michael and Malloch 1981; Ng et al 1976; Young and Stevens 1974; Ziegler et al 1982). That antibody raised against core polysaccharide structures can reach these determinants in heterologous serotypes of smooth organisms is supported by two other lines of evidence. First, when smooth organisms are injected, several antibodies are produced which recognize different lengths of the immunodominant sugars of the core and O polysaccharides (Nixdorff and Schlecht 1972). This finding indicates that these determinants are seen by immunocompetent cells. Second, organisms growing in logarithmic phase do not have complete O polysaccharides (Goldman and Leive 1980; Munford et al 1980). During bacteremia, organisms will be rapidly proliferating. Thus antibody to core antigens should be able to interact with the endotoxin of bacteremic organisms and with endotoxin released from lysing organisms (Shenep and Morgan 1984; Shenep et al 1985). However, under in vitro conditions, *E. coli* may pack its O antigens tightly and not allow the J5 antibody to reach the common core antigen (Gigliotti et al 1985).

Because the lipid A portion of endotoxin contains most of the toxic activity of endotoxin (Galanos et al 1971), antibodies to endotoxin must be able to block either free or bacterial cell-bound lipid A. The importance of steric hindrance is suggested by preliminary data showing IgM to be superior to IgG in inhibiting endotoxin toxicity (Ziegler et al 1982), an observation which supports the earlier findings of Kim and Watson (1965) that 19S antibodies are protective. Also endotoxin–anti-endotoxin complexes show little toxicity when large lactices are formed, but retain some toxicity when these lactices are small (Ogata and Kanamori 1978). These data suggest that IgM antibodies to rough organisms or against endotoxins derived from these organisms can offer protection by steric hindrance. Nevertheless, IgG fractions of anti-endotoxin produced against *E. coli* J5 and *S. minnesota* R595 protect animals (Braude and Douglas 1972; Braude et al 1973). IgG makes up the bulk of the antibodies found in patients convalescing from gram-negative bacteremia or in normal human serum (Young et al 1975a, 1975b). Thus either IgG or IgM antibodies to 'core' glycolipid may confer protection, but IgM antibodies may sterically hinder the endotoxin more efficiently. In contrast, O-specific antibodies, which give protection against a single serotype (Issekutz et al 1982; Nowotny et al 1965; Radvany et al 1966; Tate et al 1966) are predominantly of the IgG class (Young et al 1975b; Zinner and McCabe 1976).

Since lipid A contains the toxic principle of endotoxin, antibodies raised against lipid A might be expected to be most broadly protective. In fact, lipid A vaccines have not proved effective in preventing lethal bacteremia (Bruins et al 1977; Hodgin and Dreivs 1976; McCabe et al 1977; Mullan et al 1974). In addition, absorption of Re antiserum with lipid A did not diminish the protective activity of the antisera (McCabe et al 1977). Lipid A antiserum is not antipyretic when incubated with lipid A or injected intravenously before lipid A challenge (Rietschel and Galanos 1977). Some protection was shown in that lipid A antiserum prevented the local Shwartzman reaction (Rietschel and Galanos 1977). Lipid A antiserum does prevent endotoxin-induced abortion, but the abortive and lethal toxic effects are mediated by different biological pathways (Rioux-Darrieulat et al 1978). Thus a lipid A vaccine for the neutralization of endotoxin does not show much promise. This demonstrates that loss of just the 3 KDOs (3-deoxy-D-*manno*-2-octulosonic acid; 2-keto-3-deoxyoctonate) is sufficient to change the Re endotoxin into an inefficient vaccine. That KDO is exposed to antibody attack while the lipid A portion of endotoxin is hidden in smooth organisms, is supported by antibody absorption studies (Johns et al 1977).

3.2. Complement-mediated bacterial lysis

Another mechanism by which anti-endotoxin antibody might protect the host is enhancement of complement-mediated bacterial lysis. O-specific IgM antibodies are most efficient at mediating the lysis of serum-sensitive gram-negative bacilli (Bjornson and Michael 1970; Michael et al 1962; Muschel and Treffers 1956). However, most bacteremic strains of gram-negative rods are serum-resistant (Roantree and Rantz 1960). Studies of anti-core antibodies demonstrate no enhancement of complement-mediated bacteriolysis (Braude et al 1973; Chedid et al 1968; Ziegler et al 1973b), suggesting that complement-mediated lysis is not an important mechanism of protection afforded by core vaccines.

3.3. Opsonic activity

Even though the core antibody does not seem to mediate complement lysis of these bacilli, these same substances can be effective opsonins. O-specific antibody acts as an efficient opsonin (Benacerraf et al 1959b; Fisher 1977; Harvath et al 1976; Pennington and Menkes 1981; Tate et al 1966; Van Dijk et al 1981; Young et al 1975b; Ziegler et al 1973b) and also increases intracellular killing within fixed macrophages (Rowley 1958). Hepatic and splenic macrophages are the most important phagocytes in clearing bacteremia (Benacerraf et al 1959a, 1959b). Indeed, granulocytopenic dogs, immunized with the Re mutant of *S. minnesota*, cleared *E. coli* or *S. marcescens* as well as normal dogs given the Re vaccine (Young et al 1975b). Although some authors have found enhanced clearance of gram-negative rods in the presence of anti-core antibodies (Bruins et al 1977; Dunn and Ferguson 1982; Ziegler et al 1973b, 1975), the opsonic activity is 100-fold less effective when direct comparisons are made with O-specific antibody (Young et al 1975b). One possible explanation for these divergent findings is that anti-core antibodies enhance clearance of serum-sensitive, but not serum-resistant, organisms (Young et al 1975b). Serum resistance was not specifically studied by most investigators; however, most investigators used bacteremic strains of which the majority are serum-resistant (Roantree and Rantz 1960). Differences in serum sensitivity therefore probably do not fully account for the reported differences in opsonic activity of anti-core antibodies. In any event, even when no enhanced clearance can be demonstrated, the lethal, hemodynamic sequelae of gram-negative rod bacteremia are prevented by core-specific antibodies (Young and Stevens 1977). Thus when all the potentially protective mechanisms are considered, the protective effect of core vaccination seems to be mediated primarily through antitoxin activity.

3.4. Nonspecific macrophage stimulation

Active immunization has the potential of stimulating host phagocytic cells. Because this potentially protective mechanism is so intrinsically tied to animal studies, this is discussed in the next section under bacteremia.

4. PROTECTIVE STUDIES IN ANIMALS

Since endotoxin and gram-negative sepsis cause a wide variety of effects, the protective actions of passive antibody and active vaccination with rough organisms will be considered against the major toxic effects of endotoxin: bacteremia death, endotoxemia/shock, fever, local Shwartzman reaction, and diffuse intravascular coagulopathy.

4.1. Bacteremia

Core-specific antibodies, raised actively or transferred passively, have been effective in reducing the mortality in animals challenged with antigenically homologous and heterologous organisms (Braude et al 1977; Bruins et al 1977; Chedid et al 1968; Dunn and Ferguson 1982; Marks et al 1982; McCabe 1972; McCabe et al 1973a, 1977; Young and Stevens 1977; Young et al 1975b; Ziegler et al 1973b). Both passive and active protection studies are necessary to establish that antibody affords protection. Since bacteria, yeast and endotoxins can increase the number and phagocytic activity of fixed macrophages (Benacerraf and Sebestyen 1957; Boehme and Dubos 1958; Dubos and Schoedler 1957), vaccination with rough mutants might mediate protection as a nonspecific stimulator of macrophage activity rather than as an inducer of antibody. Dogs immunized with endotoxin from serotype 6 *Pseudomonas aeruginosa* (smooth organisms) were protected against intravenous challenge with 10^7 viable serotype 6 *P. aeruginosa*, whereas passive transfer of antibody failed (Harvath et al 1976). This protection was serospecific and suggested that cell-mediated immunity might be important. In contrast, human antisera, developed against a heptavalent mixture of *P. aeruginosa* smooth lipopolysaccharides, protected mice from bacteremia and death following intraperitoneal challenge with *P. aeruginosa* (Fisher 1977). *P. aeruginosa* vaccination did not provide heterologous protection (Fisher 1977; McCabe et al 1977). However, when organisms other than *P. aeruginosa* were used for vaccination, many of the studies noted above did demonstrate passive protection against challenge with organisms of heterologous serotypes (Braude et al 1977; Chedid et al 1968; Dunn and Ferguson 1982; Marks et al 1982; McCabe 1972; McCabe et al 1973a; Young and Stevens 1977; Young et al 1975b; Ziegler et al 1973b), thus strongly implicating antibody as protective. Two groups of workers have been unable to demonstrate passive protection against serotypically heterologous challenge organisms (Greisman et al 1978; Ng et al 1976; Van Dijk et al 1981). The antisera were raised in rabbits with the same rough mutants (Re mutant of *S. minnesota* and J5 mutant of *E. coli*) as other workers had used to demonstrate protection. A major problem with these three studies was that mice were used. Intravenous challenge of mice with gram-negative rods demonstrates a very steep dose–response curve, such that the difference between a lethal and sublethal dose is less than 1 order of magnitude. Also, rabbit serum may be toxic to mice (Braude 1980). Although passive transfer of protection has been demonstrated in mice (Chedid et al 1968; Fisher 1977; McCabe et al 1973a; Young et al 1975b) conflicting results would be expected with the murine model when the lethal dose curves are so steep. Overall, the data are quite compelling that antibody to rough mutants provides protection against lethal bacteremia in animals.

4.2. Protection from endotoxin shock

Studies in rabbits, rats, and mice have demonstrated that rough mutants provide an effective vaccine against endotoxin challenge (Brown et al 1971; Davis et al 1969; Freedman 1959; Kim and Watson 1966; Tate et al 1966; Ziegler et al 1973a). All of these investigators used passive transfer of antisera to effect protection. Anti-core-specific antibodies provided protection against endotoxins derived from smooth organisms of a species different from the rough strain. Of note, this heterologous protection was just as effective as protection from O-antigen-specific antibody.

4.3. Protection from fever

Early studies by Kim and Watson (1965) suggested that animals could be rendered 'tolerant' to endotoxin-induced fever by passive transfer of antibody (Greisman and Hornick 1973; Milner 1973; see also Chapter 14, Volume 2 of this series for a review of febrile patterns of tolerance). They suggested that O-specific antibodies did not fully account for the protective effect of passively transferred sera because tolerance to heterologous endotoxins was also demonstrated. Other workers questioned the role of antibody-mediated febrile tolerance in passively transferred sera (Wolff et al 1965). In a review by Greisman and Hornick (1973), a very strong case for antibody mediation of 'late' tolerance (tolerance to endotoxic challenge after 7 days) was made. Administration of antibodies with O antigen specificity results in tolerance. Although protection which was mediated through core-specific antibodies was mentioned, evidence was not presented on the protective effects of core-specific antibodies (Greisman and Hornick 1973). Braude (1980) reported that antiserum against the core region of endotoxin prevented the pyrogenicity of endotoxins derived from heterologous species, but again data were not presented nor were there references to studies. Antibody to endotoxin thus clearly seems to mediate the late tolerance, but data on prevention of the pyrogenic response to heterologous endotoxins by antibodies to core antigens are not yet available.

4.4. Protection from lethal gram-negative pneumonia

Gram-negative pneumonias have 20–50% mortality rates (McCabe and Jackson 1962a; Stevens et al 1974). *P. aeruginosa* pneumonias have an overall mortality of 40% and a mortality of more than 70% in patients with another rapidly fatal underlying disease (McCabe and Jackson 1962b; Pollack and Young 1979; Stevens et al 1974). Bacteremias often accompany the fatal pneumonias.

Since vaccination with *E. coli* J5 (Braude et al 1977; Ziegler et al 1975) or with smooth lipopolysaccharides from *P. aeruginosa* (Fisher 1977; Harvath et al 1976) protected against *P. aeruginosa* bacteremia and death, two groups were encouraged to study the effects of vaccination on gram-negative pneumonias (LaForce 1977; Pennington 1979). Mice were immunized by aerosal administration of killed Re *S. minnesota* and then challenged with aerosolized *S. marcescens, Enterobacter cloacae*, or *P. aeruginosa* (LaForce 1977). Immunization protected the mice against *S. marcescens* and *E. cloacae* challenge by enhancing lung bactericidal activity but did not protect them from *P. aeruginosa*. Further studies showed that protection against *S. marcescens* and *E. cloacae* challenge was due to proliferation and activation of lung macrophages and recruitment of polymorphonuclear leukocytes (LaForce et al 1980). In vivo or in vitro activation of alveolar macrophages did not enhance bactericidal activity against *P. aeruginosa* (LaForce et al 1980; Reynolds et al 1974). Moreover, systemic vaccination with Re *S. minnesota* did not significantly enhance the pulmonary bactericidal activity of mice challenged with *S. marcescens* despite high levels of antibody in the bronchial fluid (LaForce et al 1980). Similarly, *P. aeruginosa* immunization has demonstrated no heterologous protection (McCabe et al 1977). In contrast, systemic immunization with a heptavalent lipopolysaccharide *P.*

aeruginosa vaccine protected guinea pigs from aspiration-induced *P. aeruginosa* pneumonia (Pennington 1979; Pennington and Miler 1979). This protection was mediated by IgM opsonins which enhanced alveolar macrophage phagocytosis in the absence of active complement (Pennington and Kuchmy 1980). Neither parenteral nor lung vaccination with living *P. aeruginosa* activated alveolar macrophages (Pennington and Kuchmy 1980). When type-specific vaccination was directly compared with cross-protective vaccination in a guinea pig *P. aeruginosa* pneumonia model, the *P. aeruginosa* lipopolysaccharide vaccine gave 81% survival, the J5 *E. coli* vaccine gave 42% survival, and control animals showed 15% survival (Pennington and Menkes 1981). Although the J5 vaccine had tripled the survival rate, it was still only half as effective as the type-specific vaccine. When *E. coli* or *K. pneumoniae* were used as the challenge organisms, recipients of the J5 vaccine demonstrated no protection (Pennington and Menkes 1981). This result is quite striking since *E. coli* and *K. pneumoniae* are more closely related to the J5 *E. coli* than *P. aeruginosa*. Thus rough mutant-derived vaccines given parenterally have not been protective against gram-negative pneumonia except for the modest protection against *P. aeruginosa*. In contrast, the heptavalent lipopolysaccharide pseudomonas vaccine is protective against *P. aeruginosa* challenge.

4.5. Prevention of local and systemic Shwartzman reaction

Disseminated intravascular coagulation in rabbits follows the second of two injections of endotoxin given 24 hours apart. This coagulopathy results in deposition of fibrin in the glomeruli which leads to bilateral renal cortical necrosis (Sanarelli-Shwartzman reaction). If the first dose of endotoxin is injected intradermally, then the next systemic injection leads to thrombosis and dermal necrosis at the local injection site (local Shwartzman reaction).

Shwartzman (1931) was the first to demonstrate that antityphoid horse serum would protect rabbits against typhoid culture filtrate-mediated coagulation. However, Cluff and Bennett (1951) were unable to prevent the local Shwartzman reaction. Braude and Douglas (1972), using J5 *E. coli* and more refined endotoxins, were able to prevent the local Shwartzman reaction in rabbits challenged with *E. coli*, *Salmonella typhimurium*, and *S. marcescens* endotoxins by passively transferring J5 antisera. Of note, J5 antibodies afforded more efficient protection than O-specific antibodies. These findings were further established (Braude 1980; Ziegler et al 1973a) with other animal models (human antisera transferred to mice), and extended to include the generalized Shwartzman reaction (Braude et al 1973).

Intravascular coagulopathy occurs with high frequency during meningococcemia. Endotoxins derived from serogroups A, B, or C of *N. meningitidis* did not produce either local or diffuse intravascular coagulation in rabbits passively protected by anti-core antibodies (Davis et al 1978). The inhibition of the local Shwartzman reaction was superior to antimeningococcal sera, even when homologous antisera-endotoxin pairs were compared to anti-core sera. This observation is quite striking since group B meningococcal capsules are not immunogenic in man.

Another model of vascular damage with dermal necrosis has been described by Issekutz et al (1982). Either live or killed *E. coli* injected intradermally into rabbits causes serious vascular damage, neutrophil infiltration, and tissue necrosis. Antibody specific to O antigens or core antigens (*E. coli* J5), but not K or H antigens, was able to prevent this damage induced by killed *E. coli*. The damage caused by live *E. coli* was not inhibited by J5-antisera but was prevented by O-specific, homologous antisera. At least some of the damage was due to neutrophil infiltration. The J5-antisera were as effective as O-specific antisera in preventing the endotoxin-mediated damage from these organisms (Issekutz et al 1982). Thus core-specific antibodies are highly effective at preventing endotoxin-induced coagulopathy against a serologically heterogeneous group of organisms, but may not prevent some of the damage caused by neutrophil infiltration in response to chemotactic factors released by live *E. coli*.

5. PROTECTIVE STUDIES IN MAN

5.1. Protection from bacteremia, shock, and death

Gram-negative rod infections in man are accompanied by many of the same findings as in animals. McCabe et al (1972) reported that the frequency of shock or death in gram-negative bacteremia was significantly reduced in patients with antibody titers of 1:80 or greater to the Re antigen of *S. minnesota*. Higher titers were associated with less shock and death. Moreover, the protection was even more striking in patients who had ultimately fatal underlying diseases (McCabe et al 1972). Antibody titers to O and common antigens did not correlate with reduced incidence of shock or death (McCabe

172

et al 1972, 1973a). However, further studies using the same organisms and sera revealed that high titers of IgG, O-specific antibodies directed against the O-antigen of the patient's infecting organism were protective (McCabe et al 1977; Zinner and McCabe 1976). These O-specific antibodies afforded no protection against heterologous serotypes. Ziegler et al (1978) found that passive transfer of J5 antisera resulted in a two-fold reduction in death in 136 bacteremic patients as compared to patients that received the preimmunization control human sera (26% control, 14% antiserum). In patients with severe bacteremic shock, the death rate fell from 89% in control patients to 29% in patients receiving antisera. Similar results were found as the trial was extended to a larger group of patients (212 bacteremic cases) (Ziegler et al 1982). Passive transfer of anti-J5-antibody reduced the incidence of shock and death in a large (242 patients), randomized trial in surgical patients (Baumgartner et al 1985). A wide variety of gram-negative organisms were isolated. Thus, both the Re and J5 mutants provided heterologous protection.

5.2. Reduced incidence of fever in neutropenic patients

Prophylactic administration of J5 antisera to neutropenic patients reduced the incidence of fever from 44% in controls to 18% in those treated when patients with clear-cut evidence of fungal or gram-positive bacterial infections were excluded (Wolf et al 1979). Since the antibodies persist for at least one month, prophylactic antisera were given every 3 weeks (Wolf et al 1979).

5.3. Protection from gram-negative pneumonia

In reviewing the animal studies, the ineffectiveness of rough mutant vaccines to protect against gram-negative pneumonia was noted (see section 4.4. above). In 212 patients with infections reported by Ziegler et al (1982), 30 had gram-negative pneumonia. Although a protective trend was noted, this subgroup was too small to analyze for statistical significance. Thus passive antibody to core endotoxin structures has shown some effectiveness in clinical trials. Whether active immunization will afford even greater protection remains to be determined.

5.4. Protection in *P. aeruginosa* infections

Because *P. aeruginosa* bacteremia carries mortality rates of 40–70% (McCabe and Jackson 1962a; Ziegler et al 1982), *P. aeruginosa* infections will be considered separately. Studies in patients with acute leukemia, cystic fibrosis, solid tumors, and burn patients have suggested that antibodies to *P. aeruginosa* lipopolysaccharide provide some protection from *P. aeruginosa* infections (Alexander et al 1971; Haghbin et al 1973; Pennington et al 1975; Young et al 1973). However, evidence for clear-cut clinical efficacy is lacking because some of the studies used historical rather than double-blind controls (Alexander et al 1971) and because protective titers of antibody are lost in the face of continued immunosuppression (Young et al 1973). As patients became immunosuppressed and neutropenic, the waning levels of antibody failed to protect them from sepsis.

Another variable which confounds the understanding of the role of antibody to *P. aeruginosa* endotoxin is the role of antibody to *P. aeruginosa* exotoxin A. Pollack and Young (1979) have shown that serum antibodies to exotoxin A are of equal importance in preventing *P. aeruginosa* sepsis as antibodies to lipopolysaccharide. This may be due to the fact that pseudomonas exotoxins are toxic to macrophages and neutrophils (Baltch et al 1982; Pollack and Anderson 1978). Thus, even if antibodies to *P. aeruginosa* and phagocytes are present, phagocytic bactericidal capacity may be eliminated by the effects of uninhibited exotoxins. Combining granulocyte transfusions with antibodies to lipopolysaccharide and exotoxin A may give more clear-cut protection, especially in the immunocompromised patient. Nevertheless, protection from bacteremic death was demonstrated in patients with antibodies to rough gram-negative rod mutants in spite of the fact that *P. aeruginosa* bacteremia was present in 16 of 182 patients in one study (Zinner and McCabe 1976) and in 44 of 191 patients in another study (Ziegler et al 1982). Although these results are encouraging, larger number of patients with *P. aeruginosa* bacteremia will be needed to verify a protective effect of these antibodies in *P. aeruginosa* bacteremia definitively.

6. ADVERSE EFFECTS OF ENDOTOXIN-SPECIFIC ANTIBODIES

The benefits of any vaccine must be weighed against the risks. Immune complexes are known to stimulate tissue damage and might develop as endotoxin–anti-endotoxin aggregates in immune animals. Although antibodies to endotoxins are not necessary for endotoxin toxicity (Kim and Watson 1966; McLean 1977), aggregates of endotoxin and O-specific anti-endotoxin are lethal when given intravenously but not intraperitoneally (Kim and Watson 1966). Moreover, mice which had received endotoxin antisera prepared in rats developed an anaphylactoid reaction when challenged with endotoxin (Davis et al 1969). This reaction did not occur when 19S or 7S antibodies were given separately (Brown et al 1971). When endotoxin plus endotoxin antisera plus complement interact, relatively large complexes form which are able to release vasoactive amines (Brown et al 1971). Another measure of anaphylaxis is the contraction of ileal segments following challenge test compounds. Endotoxin alone caused no contraction in nonimmune animals (McLean 1977). Passive anti-endotoxin sera or active immunization with *E. coli* O26:B6 resulted in dose-dependent contractions of ileal segments challenged with endotoxin (McLean 1977). Finally, some of the inflammatory damage in the lungs of cystic fibrosis patients is thought to follow from immune complexes containing antibodies with endotoxin specificity (Berdischewsky et al 1980). In all these reports, the antibodies most likely had specificity for the O-antigens of endotoxin.

Platelet interactions with endotoxin–antibody complex may also be detrimental to the host. Gamma globulin aggregates, when mixed with endotoxin or lipid A, result in serotonin release from platelets (Ginsberg and Henson 1978). Neither gamma globulin aggregates nor Re endotoxin alone resulted in significant serotonin release. Similarly, antibodies with O, core, or lipid A specificity, when combined with endotoxin, caused in vitro platelet aggregation (Abdelnoor et al 1980; Walker and Beasley 1980). Nevertheless, antibodies raised against J5 *E. coli* prevent the thrombocytopenia which accompanies endotoxemia and gram-negative sepsis (Braude et al 1973; Dunn and Ferguson 1982). Thus the anaphylactoid types of reactions seem to accompany antibodies with O-antigen specificity, and the platelet–immune complex interactions are prevented, rather than enhanced, by core-specific antibodies. Moreover, although adjuvants can sensitize animals to endotoxin challenge (Michalek et al 1980), vaccination with rough mutant has been protective rather than sensitizing.

7. SUMMARY AND FUTURE STUDIES

The large number of animal studies, when taken together with a restricted number of human studies, demonstrate that immunization with rough mutants of *E. coli* or *S. minnesota* has great potential for protecting patients from fever, coagulopathy, bacteremia, and shock caused by a broad spectrum of gram-negative organisms. Protection has been demonstrated for organisms ranging from meningococci and *Haemophilus influenza* to *K. pneumoniae* and *S. marcescens*. These rough mutants are not particularly effective in the setting of experimental gram-negative pneumonias, but clinical trials may yet prove their efficacy in man. The endotoxin vaccines prepared for *P. aeruginosa*, especially when active immunization is used, show protection against *P. aeruginosa* pneumonia and bacteremia in animals, but definitive studies in man are needed. One particularly interesting new approach has been the development of monoclonal antibodies against *P. aeruginosa* (Sadoff et al 1982). These monoclonal antibodies cross-react with the rough mutant and provide passive protection against heterologous organisms. The use of anti-sera with specificity to both *P. aeruginosa* endotoxin and exotoxin A may provide a combination which will be effective in treating *P. aeruginosa* pneumonia.

Future studies will be needed to establish (a) whether active or passive immunization will provide the greatest protection, (b) whether passive immunization will be of the greatest benefit in patients who are severely immunosuppressed, (c) whether one rough mutant will be superior to another, (d) whether antibodies will reach and be effective on urethral or gastrointestinal surfaces, and (e) whether other gram-negative infections such as those caused by *Neisseria gonorrhoeae* or *Bacteroides fragilis* will be favorably affected. Clearly, as we are faced with more gram-negative infections, more antibiotic resistance, and a greater proportion of severely immunocompromised patients, immunotherapy of gram-negative infections will become increasingly important.

ACKNOWLEDGEMENTS

I would like to thank Dr. Dean Manning for his review and comments on the manuscript, Ms. Ann Campoll for preparing the manuscript, and Mr. S. Scott Westly of Merck Sharp and Dohme for his assistance in literature searches.

REFERENCES

Abdelnoor AM, Kassem H, Bikhazi AB, Nowotny A (1980) Effect of gram-negative bacterial lipopolysaccharide-derived polysaccharides, glycolipids, and lipopolysaccharides on rabbit and human platelets in vitro. *Immunobiology 157*, 145-153.

Abernathy RS, Spink WW (1958) Studies with brucella endotoxin in humans. The significance of susceptibility to endotoxin in the pathogenesis of brucellosis. *J. Clin. Invest. 37*, 219-231.

Alexander JW, Fisher MW, MacMillan BG (1971) Immunological control of pseudomonas infection in burn patients. A clinical evaluation. *Arch. Surg. 102*, 31-39.

Baltch AL, Smith RP, Hammer MC, Bishop M, Lutz F (1982) The effect of *Pseudomonas aeruginosa* cytotoxin on human granulocytes. In: *Abstracts of the 22nd Interscience Conference on Antimicrobial Agents Chemotherapy*, No. 747, p 197. American Society for Microbiology, Washington D.C.

Baumgartner JD, Glauser MP, McCutchan JA, Ziegler E, Van Melle G, Klauber MR, Vogt M, Muehlen E, Luethy R, Chiolero R, Geroulanos S (1985) Prevention of gram-negative shock and death in surgical patients by antibody endotoxin core glucolipid. *Lancet 2*, 59-63.

Beckmann I, Subbaiah TV, Stocker BAD (1964) Rough mutants of *Salmonella typhimurium*. (2). Serological and chemical investigations. *Nature (London) 201*, 1299-1301.

Benacerraf B, Sebestyen MM (1957) Effect of bacterial endotoxins on the reticuloendothelial system. *Fed. Proc. 16*, 860-868.

Benacerraf B, Kivy-Rosenberg E, Sebestyen MM, Zweifach BW (1959a) The effect of high doses of X-irradiation on the phagocytic, proliferative, and metabolic properties of the reticuloendothelial system. *J. Exp. Med. 110*, 49-64.

Benacerraf B, Sebestyen MM, Schlossman S (1959b) A quantitative study of the kinetics of blood clearance of P^{32}-labeled *Escherichia coli* and staphylococci by the reticuloendothelial system. *J. Exp. Med. 110*, 27-48.

Berdischewsky M, Pollack M, Young LS, Chia D, Osher AB, Barnett EV (1980) Circulating immune complexes in cystic fibrosis. *Pediatr. Res. 14*, 830-833.

Bjornson AB, Michael JG (1970) Biological activities of rabbit Immunoglobulin M and Immunoglobulin G antibodies to *Pseudomonas aeruginosa*. *Infect. Immun. 2*, 453-461.

Boehme D, Dubos RJ (1958) The effect of bacterial constituents on the resistance of mice to heterologous infection and on the activity of their reticuloendothelial system. *J. Exp. Med. 107*, 523-536.

Boivin A, Mesrobeanu L (1938) Recherches sur les antigènes somatiques et sur les endotoxins des bactéries. IV. Sur l'action antiendotoxique de l'anticorps O. *Rev. Immunol. 4*, 40-56.

Braude AE (1980) Endotoxic immunity. *Adv. Intern. Med. 26*, 427-445.

Braude AI, Douglas H (1972) Passive immunization against the local Shwartzman reaction. *J. Immunol. 108*, 505-512.

Braude AI, Douglas H, Davis CE (1973) Treatment and prevention of intravascular coagulation with antiserum to endotoxin. *J. Infect. Dis. 128*, S157-S164.

Braude AI, Ziegler EJ, Douglas H, McCutchen JA (1977) Antibody to cell wall glycolipid of gram-negative bacteria: induction of immunity to bacteremia and endotoxemia. *J. Infect. Dis. 136*, S167-S173.

Brown KR, Douglas H, Braude AI (1971) Prevention of death from endotoxin with antiserum. II. Elimination of the risk of anaphylaxis to endotoxin. *J. Immunol. 106*, 324-333.

Bruins SC, Stumacher R, Johns MA, McCabe WR (1977) Immunization with R mutants of *Salmonella minnesota*. III. Comparison of the protective effect of immunization with lipid A and the Re mutant. *Infect. Immun. 17*, 16-20.

Bryant RE, Hood AF, Hood CE, Koenig MG (1971) Factors affecting mortality of gram-negative rod bacteremia. *Arch. Intern. Med. 127*, 120-128.

Burke JF (1961) The effective period of preventive antibiotic action in experimental incisions and dermal lesions. *Surgery 50*, 161-169.

Chedid L, Parant M, Parant F, Boyer F (1968) A proposed mechanism for natural immunity of enterobacterial pathogens. *J. Immunol. 100*, 292-301.

Christy JH (1971) Treatment of gram-negative shock. *Am. J. Med. 50*, 77-88.

Cluff LE, Bennett IL (1951) Acquired resistance to the Shwartzman phenomenon. *Proc. Soc. Exp. Biol. Med. 77*, 461-472.

Creech HJ, Hamilton MA, Nishimura ET, Hankwitz RF Jr (1948) The influence of antibody-containing fractions on the lethal and tumor-necrotizing actions of polysaccharides from *Serratia marcescens (Bacillus prodigiosus)*. *Cancer Res. 8*, 330-338.

Davis C, Brown K, Douglas H, Tate W, Braude A (1969) Prevention of death from endotoxin with antisera. I. The risk of fatal anaphylaxis to endotoxin. *J. Immunol. 102*, 563-572.

Davis CE, Ziegler EJ, Arnold KF (1978) Neutralization of meningococcal endotoxin by antibody to core glycolipid. *J. Exp. Med. 147*, 1007-1017.

Dubos RJ, Schoedler RW (1957) Effects of cellular constituents of mycobacteria on the resistance of mice to heterologous infections. I. Protective effects. *J. Exp. Med. 106*, 703-718.

Dunn D, Ferguson R (1982) Immunotherapy of gram-negative bacterial sepsis: enhanced survival in a guinea pig model by use of rabbit antiserum to *Escherichia coli* J-5. *Surgery 92*, 212-219.

Edwards PR, Ewing WH (1972) *Identification of Enterobacteriaceae, 3rd Edition*. Burgess, Minneapolis.

Fisher MW (1977) A polyvalent human γ-globulin immune to *Pseudomonas aeruginosa*: passive protection of mice against lethal infection. *J. Infect. Dis. 136 (Suppl)*, S181-S185.

Freedman H (1959) Passive transfer of protection against lethality of homologous and heterologous endotoxins. *Proc. Soc. Exp. Biol. Med. 102*, 504-506.

Galanos C, Lüderitz B, Westphal O (1971) Preparation and properties of antisera against the lipid-A component of bacterial lipopolysaccharides. *Eur. J. Biochem. 24*, 116-122.

Gigliotti F, Shenep JI (1985) Failure of monoclonal antibodies to core glycolipid to bind intact smooth strains of *Escherichia coli*. *J. Infect. Dis. 151*, 1005-1011.

Ginsberg MH, Henson PM (1978) Enhancement of platelet response to immune complexes and IgG aggregates by lipid A-rich bacterial lipopolysaccharides. *J. Exp. Med. 147*, 207-218.

Glew RJ, Mollering RC Jr, Burke JF (1976) Gentamicin dosage in children with extensive burns. *J. Trauma 16*, 819-825.

Goldman RC, Leive L (1980) Heterogeneity of antigenic-side-chain length in lipopolysaccharide from *Escherichia coli* 0111 and *Salmonella typhimurium* LT2. *Eur. J. Biochem. 107*, 145-153.

Goldman RC, White D, Ørskov F, Ørskov I, Rick PD, Lewis MS, Bhattacharjee AK, Leive L (1982) A surface polysaccharide of *Escherichia coli* O111 contains O-antigen and inhibits agglutination of cells by O-antiserum. *J. Bact. 151*, 1210-1221.

Greisman S, Hornick R (1973) Mechanisms of endotoxin tolerance with special references to man. *J. Infect. Dis. 128*, S265-S276.

Greisman SE, Dubuy JB, Woodward CL (1978) Experimental gram-negative bacterial sepsis: reevaluation of the ability of rough mutant antisera to protect mice. *Proc. Soc. Exp. Biol. Med. 158*, 482-490.

Haghbin M, Armstrong D, Murphy ML (1973) Controlled prospective trial of *Pseudomonas aeruginosa* vaccine in children with acute leukemia. *Cancer (Philadelphia) 32*, 761-766.

Harvath L, Anderson BR, Amirault HJ (1976) Passive immunity against Pseudomonas sepsis during granulocytopenia. *Infect. Immun. 14*, 1151-1155.

Heath EC, Mayer RM, Edstrom RD, Beaudreau CA (1966) Structure and biosynthesis of the cell wall lipopolysaccharide of *Escherichia coli*. *Ann. NY Acad. Sci. 133*, 315-330.

Hodgin L, Dreivs J (1976) Effect of active and passive immunizations with lipid A and *Salmonella minnesota* Re 595 on gram-negative infections in mice. *Infection 4*, 5-13.

Issekutz AC, Bhimji S, Bortolussi R (1982) Effect of immune serum or polymyxin B on *Escherichia coli*-induced inflammation and vascular injury. *Infect. Immun. 36*, 548-557.

Ito JI, Lyons WJ, Davis CE, Guiney DG, Braude AI (1980) Role of magnesium in the enzyme-linked immunosorbent assay for lipopolysaccharides of rough *Escherichia coli* strain J5 and Neisseria. *J. Infect. Dis. 142*, 532-537.

Johns MA, Bruins SC, McCabe WR (1977) Immunization with R mutants of *Salmonella minnesota*. II. Serological response to lipid A and the lipopolysaccharide of Re mutants. *Infect. Immun. 17*, 9-15.

Kim YB, Watson DW (1965) Modification of host responses to bacterial endotoxins. II. Passive transfer of immunity to containing 19S antibodies. *J. Exp. Med. 121*, 751-759.

Kim YB, Watson DW (1966) Role of antibodies in reactions to gram-negative bacterial endotoxins. *Ann. NY Acad. Sci. 133*, 727-745.

Kunin CM (1963) Separation, characterization and biological significance of a common antigen to enterobacteriaceae. *J. Exp. Med. 118*, 565-586.

LaForce FM (1977) Effect of aerosal immunization with Re 595 *Salmonella minnesota* on lung bactericidal activity against *Serratia marcescens*, *Enterobacter cloacae*, and *Pseudomonas aeruginosa*. *Am. Rev. Resp. Dis. 116*, 241-249.

LaForce FM, Boose DS, Mills DM (1980) Heightened lung bactericidal activity in mice after aerosol immunization with Re595 *Salmonella minnesota*: importance of cellular rather than humoral factors. *J. Infect. Dis. 142*, 421-431.

Lüderitz O, Staub A, Westphal O (1966) Immunochemistry of O and R antigens of Salmonella and related Enterobacteriaceae. *Bact. Rev. 30*, 192-231.

Mäkelä PH, Mayer H (1976) Enterobacterial common antigen. *Bact. Rev. 40*, 591-632.

Marks MI, Ziegler EJ, Douglas H, Corbeil LB, Braude AI (1982) Induction of immunity against lethal *Haemophilus influenzae* type b infection by *Escherichia coli* core lipopolysaccharide. *J. Clin. Invest. 69*, 742-749.

McCabe WR (1972) Immunization with R mutants of *S. minnesota*. I. Protection against challenge with heterologous gram-negative bacilli. *J. Immunol. 108*, 601-610.

McCabe WR, Greely A (1973) Common enterobacterial antigen. II. Effect of immunization on challenge with heterologous bacilli. *Infect. Immun. 7*, 386-392.

McCabe WR, Jackson GG (1962a) Gram-negative bacteremia. I. Etiology and ecology. *Arch. Intern. Med. 110*, 847-855.

McCabe WR, Jackson GG (1962b) Gram-negative bacteremia. II. Clinical, laboratory, and therapeutic observations. *Arch. Intern. Med. 110*, 856-867.

McCabe WR, Kreger BE, Johns M (1972) Type-specific and cross-reactive antibodies in gram-negative bacteremia. *N. Engl. J. Med. 287*, 261-267.

McCabe WR, Greely A, DiGenio T, Johns MA (1973a) Humoral immunity to type-specific and cross-reactive antigens of gram-negative bacilli. *J. Infect. Dis. 128*, S284-S289.

McCabe WR, Johns M, DiGenio T (1973b) Common enterobacterial antigen. III. Initial titers and antibody responses in bacteremia caused by gram-negative bacilli.

Infect. Immun. 7, 393-397.

McCabe WR, Bruins SC, Craven DE, Johns M (1977) Cross-reactive antigens: their potential for immunization-induced immunity to gram-negative bacteria. *J. Infect. Dis. 136*, S161-S166.

McGowan JE Jr, Barnes MW, Finland M (1975) Bacteremia at Boston City Hospital: occurrence and mortality during 12 selected years (1935-1972), with special reference to hospital-acquired cases. *J. Infect. Dis. 132*, 316-335.

McLean AJ (1977) Anaphylactic reactions to endotoxin guinea-pig tissues: relationship to endotoxin toxicity. *Br. J. Pharmacol. 60*, 369-373.

Michael JG, Malloch I (1981) Immune response to parenteral and rough mutant strains of *Salmonella minnesota*. *Infect. Immun. 33*, 784-787.

Michael JG, Whitby JL, Landy M (1962) Studies on natural antibodies to gram-negative bacteria. *J. Exp. Med. 115*, 131-146.

Michalek SM, Moore RN, McGhee JR, Rosenstreich DL, Mergenhagen SE (1980) The primary role of lymphorecticular cells in the mediation of host response to bacterial endotoxin. *J. Infect. Dis. 141*, 55-63.

Milner K (1973) Patterns of tolerance to endotoxin. *J. Infect. Dis. 128*, S237-S245.

Morgan HR (1941) Immunological properties of an antigenic material isolated from *Eberthella typhosa*. *J. Immunol. 41*, 161-180.

Mullan WA, Newsome PM, Cunnington PG, Palmer GH, Wilson ME (1974) Protection against gram-negative infections with antiserum to lipid A from *Salmonella minnesota* R595. *Infect. Immun. 10*, 1195-1201.

Munford RS, Hall CL, Rick PD (1980) Size heterogeneity of *Salmonella typhimurium* lipopolysaccharides in outer membranes and culture supernatant membrane fragments. *J. Bacteriol. 144*, 630-640.

Muschel LH, Treffers HP (1956) Quantitative studies on the bactericidal actions of serum and complement. I. A rapid photometric growth assay for bactericidal activity. *J. Immunol. 76*, 1-10.

Myerowitz RL, Medeiros AA, O'Brien TF (1971) Recent experience with bacillemia due to gram-negative organisms. *J. Infect. Dis. 124*, 239-246.

Ng A, Chen CLH, Chang CM, Nowotny A (1976) Relationship of structure to function in bacterial endotoxins: serologically cross-reactive components and their effect on protection of mice against some gram-negative infections. *J. Gen. Microbiol. 94*, 107-116.

Nixdorff KK, Schlecht S (1972) Heterogeneity of the haemagglutinin responses to *Salmonella minnesota* R-antigens in rabbits. *J. Gen. Microbiol. 71*, 425-440.

Nowotny A, Radvany R, Neale NL (1965) Neutralization of toxic bacterial O-antigens with O-antibodies while maintaining their stimulus on non-specific resistance. *Life Sci. 4*, 1107-1114.

Ogata S, Kanamori M (1978) Effects of homologous O-antibody on host responses to lipopolysaccharide from *Yersinia enterocolitica*: neutralization of its pyrogenicity. *Microbiol. Immunol. 22*, 485-494.

Ørskov F, Ørskov I (1975) Escherichia O:H serotypes isolated from human blood. *Acta Pathol. Microbiol. Scand. (B) 83*, 595-607.

Osborn MJ (1966) Biosynthesis and structure of the core region of the lipopolysaccharide in *Salmonella typhimurium. Ann. NY Acad. Sci. 133*, 375-383.

Pennington JE (1979) Lipopolysaccharide Pseudomonas vaccine: efficacy against pulmonary infection with *Pseudomonas aeruginosa. J. Infect. Dis. 140*, 73-80.

Pennington JE, Kuchny D (1980) Mechanism for pulmonary protection by lipopolysaccharide Pseudomonas vaccine. *J. Infect. Dis. 142*, 191-198.

Pennington JE, Menkes E (1981) Type-specific vs. cross-protection vaccination for gram-negative bacterial pneumonia. *J. Infect. Dis. 144*, 599-603.

Pennington JE, Miler JJ (1979) Evaluation of a new polyvalent Pseudomonas vaccine in respiratory infections. *Infect. Immun. 25*, 1029-1034.

Pennington JE, Reynolds HY, Wood RE, Robinson RA, Levine AS (1975) Use of a *Pseudomonas aeruginosa* vaccine in patients with acute leukemia and cystic fibrosis. *Am. J. Med. 58*, 629-636.

Pfeiffer R, Kolle W (1896) Uber die specifische Immunutätsreaction der Typhus-Bacillen. *Z. Hyg. Infektionskr. 21*, 203-214.

Pollack M, Anderson SE (1978) Toxicity of *Pseudomonas aeruginosa* exotoxin A for human macrophages. *Infect. Immun. 19*, 1092-1096.

Pollack M, Young LS (1979) Protective activity of antibodies to exotoxin A and lipopolysaccharide at onset of *Pseudomonas aeruginosa* septicemia in man. *J. Clin. Invest. 63*, 276-286.

Radvany R, Neale N. Nowotny A (1966) Relation of structure to function in bacterial O-antigens. VI. Neutralization of endotoxin O-antigens by homologous O-antibody. *Ann. NY Acad. Sci. 133*, 763-786.

Reynolds HY, Thompson RE, Devlin GH (1974) Development of cellular and humoral immunity in the respiratory tract of rabbits to Pseudomonas lipopolysaccharide. *J. Clin. Invest. 53*, 1351-1358.

Rietschel ETh, Galanos C (1977) Lipid A antiserum-mediated protection against lipopolysaccharide and lipid A-induced fever and skin necrosis. *Infect. Immun. 15*, 34-49.

Rioux-Darrieulat F, Parant M, Chedid L (1978) Prevention of endotoxin-induced abortion by treatment of mice with antisera. *J. Infect. Dis. 137*, 7-13.

Roantree RJ, Rantz LA (1960) A study of the relationship of the normal bactericidal activity of human serum to bacterial infection. *J. Clin. Invest. 39*, 72-81.

Rogers DE (1959) The changing pattern of life-threatening microbial disease. *N. Engl. J. Med. 261*, 677-683.

Rowley D (1958) Bactericidal activity of macrophages in vitro against *Escherichia coli. Nature (London) 181*, 1738-1739.

Sadoff J, Sidberry H, Seid R, Futrovsky S (1982) Monoclonal antibodies against

Pseudomonas aeruginosa lipopolysaccharide that protect in vivo. In: *Abstracts of the 22nd Interscience Conference on Antimicrobial Agents Chemotherapy*, No. 253, p. 110. American Society for Microbiology, Washington DC.

Shenep JL, Morgan KA (1984) Kinetics of endotoxin release during antibiotic therapy for experimental gram-negative bacterial sepsis. *J. Infect. Dis. 150*, 380-388.

Shenep JL, Barton RP, Morgan KA (1985) Role of antibiotic class in the rate of liberation of endotoxin during therapy for experimental gram-negative bacterial sepsis. *J. Infect. Dis. 151*, 1012-1018.

Shwartzman G (1931) Phenomenon of local skin reactivity to bacterial filtrates: passive immunity to reacting factors. *J. Exp. Med. 54*, 1-10.

Stevens RM, Teres D, Skillman JJ, Feingold DS (1974) Pneumonia in an intensive care unit: a 30-month experience. *Arch. Intern. Med. 134*, 106-111.

Subbaiah TV, Stocker BAD (1964) Rough mutants of *Salmonella typhimurium. Nature (London) 201*, 1298-1299.

Tate WJ, Douglas H. Braude AE (1966) Protection against lethality of *E. coli* endotoxin with 'O' antiserum. *Ann. NY Acad. Sci. 133*, 746-762.

Van Dijk WC, Verbrugh HA, Van Erne-VanDerTol ME, Peters R, Verhoef J (1981) *Escherichia coli* antibodies in opsonisation and protection against infection. *J. Med. Microbiol. 14*, 381-389.

Waisbern BA (1951) Bacteremia due to gram-negative bacilli other than Salmonella. A clinical and therapeutic study. *Arch. Intern. Med. 88*, 467-494.

Walker RI, Beasley WJ (1980) Evidence that antibodies to polysaccharide alter platelet responses to endotoxin in tolerant rabbits. *Can. J. Microbiol. 26*, 1241-1246.

Watson DW, Kim YB (1963) Modification of host responses to bacterial endotoxins. I. Specificity of pyrogenic tolerance and role of hypersensitivity in pyrogenicity, lethality, and skin reactivity. *J. Exp. Med. 118*, 425-446.

Weil AJ, Gall LS, Wider S (1939) Progress in the study of the typhoid bacillus. *Arch. Pathol. 28*, 71-89.

Wolf JL, McCutchan JA, Ziegler EJ, Braude AI (1979) Prophylactic antibody to core lipopolysaccharide in neutropenia. In: Nelson JD, Grossi C (Eds), *Proceedings of the 11th International Congress on Chemotherapy and 19th Interscience Conference on Antimicrobial Agents and Chemotherapy*, pp 1439-1441. American Society for Microbiology, Washington DC.

Wolff SM, Mulholland JH, Ward SB, Rubenstein M, Mott PD (1965) Effect of 6-mercaptopurine on endotoxin tolerance. *J. Clin. Invest. 44*, 1402-1409.

Young LS (1979) Gram-negative sepsis. In: Mandell GL, Douglas RG Jr, Bennett JE (Eds), *Principle and Practices of Infectious Diseases*, pp 571-608. John Wiley and Sons, New York.

Young LS, Stevens PR (1974) Precipitating antibody against core glycolipid of Enterobacteriaceae. *Experientia 30*, 192-193.

Young LS, Stevens PR (1977) Cross-protective immunity to gram-negative bacilli: studies with core glycolipid of *Salmonella minnesota* and antigens of *Streptococcus pneumoniae. J. Infect. Dis. 136 (Suppl)*, S174-S180.

Young LS, Meyer RD, Armstrong D (1973) *Pseudomonas aeruginosa* vaccine in cancer patients. *Ann. Intern. Med. 79*, 518-527.

Young LS, Hoffman KR, Stevens P (1975a) 'Core' glycolipid of Enterobacteriaceae: immunofluorescent detection of antigen and antibody. *Proc. Soc. Exp. Biol. Med. 149*, 389-396.

Young LS, Stevens P, Ingram J (1975b) Functional role of antibody against 'core' glycolipid of Enterobacteriaceae. *J. Clin. Invest. 56*, 850-861.

Ziegler E, Douglas H, Braude A (1973a) Human antiserum for prevention of the local Shwartzman reaction and death from bacterial lipopolysaccharides. *J. Clin. Invest. 52*, 3236-3238.

Ziegler EJ, Douglas H, Sherman JE, Davis CE, Braude AI (1973b) Treatment of *E. coli* and *Klebsiella* bacteremia in agranulocytic animals with antiserum to a UDP-GAL-epimerase-deficient mutant. *J. Immunol. 111*, 433-438.

Ziegler EJ, McCutchan JA, Douglas H, Braude AI (1975) Prevention of lethal Pseudomonas bacteremia with epimerase-dificient *E. coli* antiserum. *Trans. Assoc. Am. Phys. 88*, 101-108.

Ziegler EJ, McCutchan JA, Braude AI (1978) Clinical trial of core glycolipid antibody in gram-negative bacteremia. *Trans. Assoc. Am. Phys. 91*, 253-258.

Ziegler EJ, McCutchan JA, Fierer J, Glauser MP, Sadoff JC, Douglas H, Braude AI (1982) Treatment of gram-negative bacteremia and shock with human antiserum to a mutant *Escherichia coli. N. Engl. J. Med. 307*, 1225-1230.

Zinner SH, McCabe WR (1976) Effects of IgM and IgG antibody in patients with bacteremia due to gram-negative bacilli. *J. Infect. Dis. 133*, 37-45.

Handbook of Endotoxin, Vol. 4: Clinical Aspects of Endotoxin Shock
R.A. Proctor, editor
© Elsevier Science Publishers B.V., 1986

CHAPTER 6

Cardiopulmonary dysfunction from sepsis: diagnosis and treatment*

ROBERT H. DEMLING

1. INTRODUCTION

Endotoxin shock is a term considered to be synonymous with bacterial or septic shock. It has already been pointed out that endotoxin and sepsis injury are not necessarily synonymous. However, many of the major organ dysfunctions caused by endotoxin or bacteria are very similar (Blain et al 1970). In this chapter, I will discuss the treatment of cardiopulmonary injury from sepsis. I will try to specify when indicated whether research findings stem from endotoxin studies or from studies on bacterial infection.

The recognition of sepsis injury as a continuum of disease with shock as the end stage has allowed us to recognize the symptoms well before complete cardiopulmonary collapse and to start therapy. The shock state is becoming less frequent because of the earlier initiation of fluid therapy. However, organ failure from sepsis remains a major cause of mortality in the critically ill or traumatized patient. Blaisdell (1981) reported an extremely low incidence of septic shock in post-trauma intensive care unit patients, but a significant mortality rate from sepsis-induced multiorgan failure. The maintenance of blood volume and early rapid restoration of sepsis-induced hypovolemia eliminated the 'shock state' but not the progressive cardiopulmonary failure. I will, therefore, not confine the discussion of treatment modalities simply to the shock state.

Although sepsis (endotoxemia) affects all organ systems to some degree, much of the injury is due to the secondary effects of cardiopulmonary failure and the ensuing hypoperfusion and hypoxia. I will, therefore, concentrate primarily on the cardiovascular and pulmonary abnormalities in this chapter.

* Work presented in this chapter was supported by National Institutes of Health grants HL 30068 and GM 31662.

2. PULMONARY DYSFUNCTION

The lung is clearly a target organ for endotoxin or sepsis injury. The predominant cause of the adult respiratory distress syndrome (ARDS) is sepsis either by itself or more commonly in association with an episode of tissue trauma. It has been reported that over 20% of patients who have a septic episode will develop lung injury (Kaplan et al 1979). Trauma patients who develop sepsis have a nearly 40% incidence of lung injury (Finley et al 1975) and in one-fifth of these patients, a respiratory dysfunction was the first symptom of a distant focus of sepsis (Vito et al 1974). This form of indirect lung injury has been reported to occur nearly 100 000 times a year in the United States, with a reported mortality in the range of 20–50% (Demling 1981). ARDS is an extremely complex syndrome and is best understood if related to normal physiology. I would, therefore, like to discuss first the basic alterations in the lung that result in respiratory failure.

3. PULMONARY MEMBRANE PHYSIOLOGY

There are two membrane systems within the lung, the microvascular membrane separating the intravascular and interstitial spaces and the alveolar membrane separating the interstitial and alveolar spaces (Fig. 1). Although injury to one membrane often leads to an impairment of the other, each has very specific characteristics that need to be discussed separately.

3.1. Microvascular membrane

The pulmonary microvascular membrane is made up of the endothelial cell layer separating the intravascular and interstitial spaces. In the normal lung, there is a restricted but net positive fluid and protein flow across this barrier. The normal pulmonary interstitium is, therefore, not water- or protein-free.

A number of theories have been proposed to explain microvascular membrane transport. The most popular theory, and the one best substantiated with experimental evidence, is that fluid and protein transport is a passive process occurring through channels or 'pores' in the membrane. A great number of small pores, 0.4–0.5 nm in radius, occur within the cell membrane and allow for water diffusion. Because of their small size, these pores are highly restrictive to solutes, in particular to protein. A lower number of

186

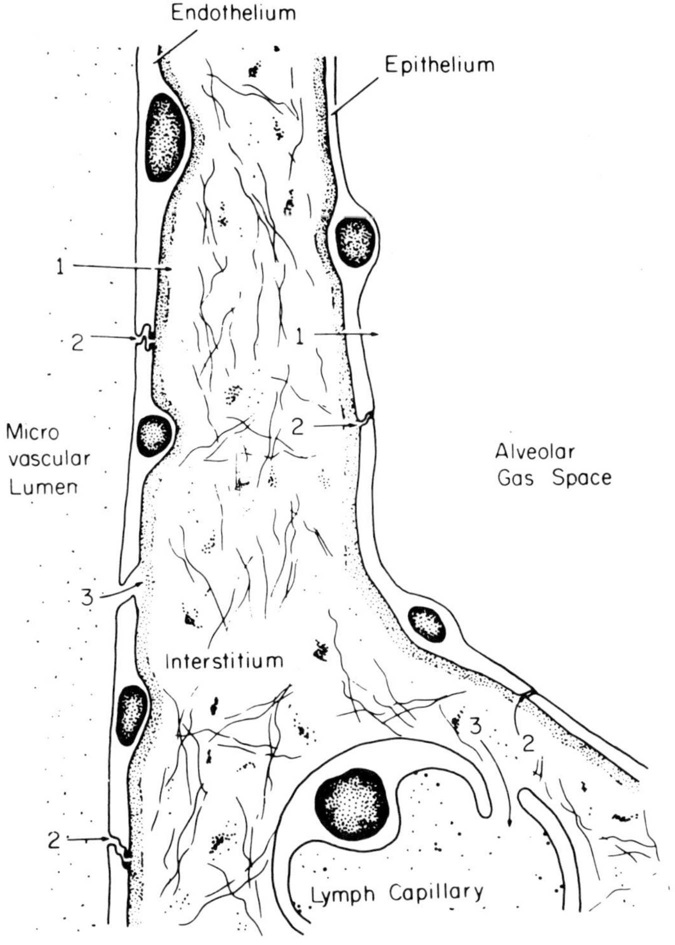

Fig. 1 *The major fluid and protein exchange features of the lung. The intravascular compartment is represented by the microvascular lumen. Fluid and protein cross into the interstitium through a number of holes or pores in the endothelial membrane. The three classes of pores shown are (1) cell membrane small pores, which water and small solutes can cross, (2) intercellular junctions, which allow proteins the size of albumin to cross, and (3) large intercellular leaks, which are found occasionally along the membrane and which allow a few larger proteins to cross. No large pores are found on the alveolar membrane, making this impermeable to protein. Excess fluid and protein are removed by the lung lymphatics. Reprinted from Staub (1974), with permission.*

larger pores, 3.5 to 6.0 nm in radius, are considered to be present at the junctions between endothelial cells, and restricted protein flow can occur. These pores are approximately the same radius as an albumin molecule. A significant amount of albumin can, therefore, cross into the pulmonary interstitium under normal conditions, with the protein content of interstitial fluid reported to be 50–80% that of plasma. Since there are large proteins such as fibrinogen present in normal lung fluid, although in small quantities, occasional large intercellular gaps of about 100 nm in radius probably also occur, which would allow a few of these large molecules to cross. The pores or the leakiness of the microvascular membrane increases from the arterial to the venular ends of the vessels. Because of this, it appears that many of the vasoactive agents reported to cause edema act on the pulmonary venules.

The amount of fluid crossing the capillary at any one time is determined by the sum of the Starling forces, or the net driving pressure, times the conductance, or ease of water transport. This is presented in the following equation: $Q_f = K_f [(Pmv - Ppmv) - \sigma(\pi mv - \pi pmv)]$ (Staub 1974), where Q_f is the net rate of fluid filtration, K_f is the coefficient describing the conductivity of water, Pmv and $Ppmv$ are the pulmonary microvascular and interstitial hydrostatic pressures, respectively, σ is the reflection coefficient, which measures the relative leakiness of the membrane to protein as compared with water, being nearly equal to one for protein and zero for small molecules, and πmv and πpmv are the microvascular and interstitial protein osmotic pressures, respectively. The importance of osmotic pressure on the filtration rate is reflected in the difference between plasma and interstitial values. An increase in water conductance generally also means an increase in leakiness to protein. A severe destruction of the membrane is not necessary for large increases in protein and fluid flow. A separation of intercellular gaps from the normal 3.5 to 10.0 nm would essentially allow free flow of albumin, but the membrane would not be visibly changed even by electron microscopy. Excess water or protein will cross the pulmonary capillary when the net driving pressure (sum of Starling forces) is increased, or when membrane permeability is increased, often a combination of these alterations is present.

Alterations in the Starling forces will be presented first. By far the most important Starling force governing fluid transport is microvascular hydrostatic or capillary pressure, Pmv. The value of Pmv is somewhere between pulmonary artery (Ppa) and left atrial (Pla) pressures. The standard equation for this calculation is $Pmv = R(Ppa - Pla) + Pla$, R being the fraction of total resistance on the venous side. In the normal lung, R is felt to be 0.4 or 40% to total resistance. An increase in Pmv resulting in 'high pressure pulmonary edema' can be produced by either a large increase in Ppa and Pla, as seen with heart failure or volume overload, or an increase in venous resistance, R. An increase in the latter probably plays a major role in ARDS and may well be the major cause of 'neurogenic pulmonary edema'. High pressure pulmonary edema is characterized by vascular congestion, extravascular red blood cells, and protein-poor edematous fluid, since protein permeability is intact. The increased red blood cells are due to focal ruptures of capillaries and red cell diapedesis. These are seen on the surface of the lung as focal hemorrhages. Except for extreme changes, decreases in plasma colloid osmotic pressure (COP) are compensated by decreases in interstitial COP resulting in no net change in the oncotic or colloid osmotic gradient. Increases in Pmv are also partly compensated by decreases in interstitial oncotic pressure. This happens because more water than protein crosses the intact membrane, diluting interstitial protein. It is estimated that approximately 50% of the effect of an increase in Pmv is buffered in this manner (Sugarman et al 1981). Severe hypoproteinemia of course eliminates this protective measure. Interstitial hydrostatic pressure is the Starling force we know least because it is difficult to measure. This pressure is felt to remain relatively constant until severe interstitial edema is present.

Increased permeability edema occurs when either the number of pores or the size of the pores or defects between endothelial cells increases. This results in a marked increase in fluid and protein crossing the membrane even with a normal Pmv (Brigham et al 1980). The plasma–interstitial oncotic gradient is now decreased since the membrane can no longer effectively hold back macromolecules. However, even with a severe alteration in permeability, there is some barrier remaining.

It cannot be overemphasized that the injured microvascular membrane is now extremely sensitive to changes in Pmv (Fig. 2). This results in rapid increases in edema for very small increases in Pmv. The microvascular injury is usually reversible if the agent producing the damage is eliminated or the membrane is stabilized (Demling 1981).

There are, therefore, two types of microvascular injury (Fig. 3) resulting in either high pressure or permeability edema. The resultant edema is initially confined to the interstitial space. As the interstitial space expands, an increase in small airway resistance can occur due to external airway compression. This and a similar increase in vascular resistance will result in ventila-

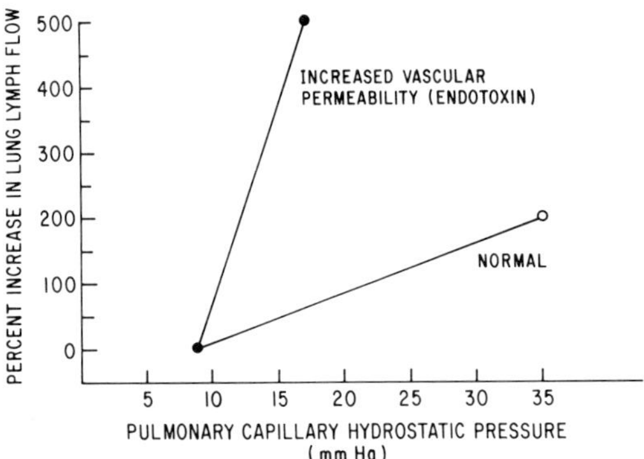

Fig. 2　*The responses of the normal and the endotoxin-injured pulmonary capillary to changes in hydrostatic pressure. Excess fluid crossing the capillary is reflected in the increased amount of lymph draining from the lung interstitium. Large increases in capillary pressure are required in the normal capillary to produce substantial increases in lung fluid leading to 'high pressure edema'. However, the capillary with increased permeability leaks large amounts of fluid in response to only minimal increases in pressure, resulting in the rapid onset of 'permeability edema'.*

tion-perfusion ($\dot{V}/\dot{Q}$) abnormalities. Further problems with gas exchange will occur if the alveolar membrane is damaged, resulting in alveolar edema.

3.2. Interstitium

The interstitial space appears to play a major role in the control of fluid and protein flux. This portion of the lung is made up of an interstitial matrix of glycosaminoglycans and protein, including tissue fibronectin. The compliance of the interstitial space appears to be a very important determinant of fluid accumulation (Comper and Laurent 1978; Greenwald and Moy 1980; Kramer et al 1981; Lee et al 1981; Vaccaro et al 1982). We have recently demonstrated that although plasma hypoproteinemia does not result in an imbalance of Starling forces because of the compensatory decrease in the interstitial value, there is still an increase in fluid flux into the interstitium (Kramer et al 1981, 1982a). This appears to be caused by an alteration in the interstitial matrix due to protein washout which decreases the viscosity of the matrix and allows easier fluid accumulation with the same balance of

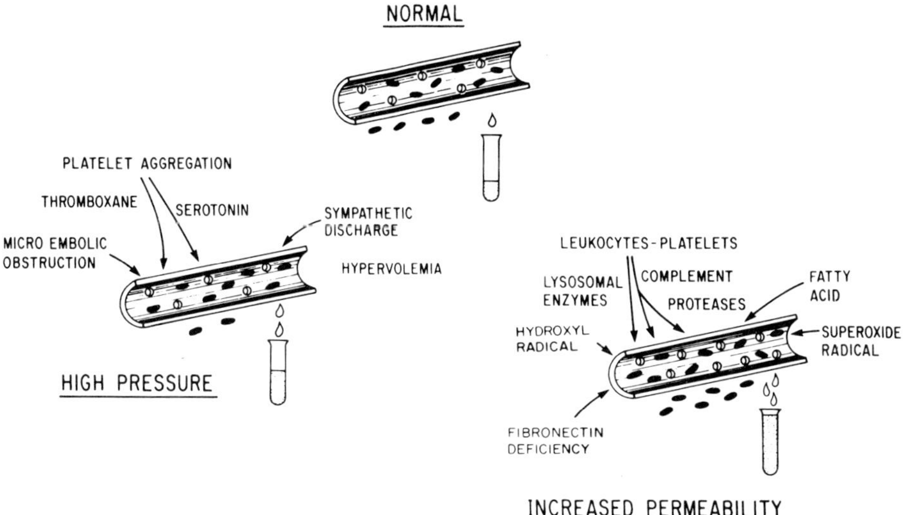

Fig. 3　*The lung capillary under normal conditions leaks protein (●) and fluid through a number of holes or pores in the membrane. Loss of membrane integrity leading to ARDS can occur if the capillary hydrostatic pressure is acutely elevated as a result of a number of insulting agents causing lung edema. Permeability is unaltered in pure 'high pressure edema', and therefore the number of pores remains constant and the interstitial protein content decreases as it becomes diluted by the excess water. An increase in permeability can also readily cause ARDS. When protein permeability is altered, the number or size of the pores increases, resulting in greater quantities of both fluid and protein crossing the membrane. 'Permeability edema' is generally of greater magnitude than 'high pressure edema'. Many of the ARDS states are a combination of these forms of injury, but one form usually predominates.*

Starling forces (Greenfield et al 1974). The lung, however, is only minimally affected by this process compared to the peripheral microcirculation where severe edema can occur by this nononcotic effect of hypoproteinemia.

Alterations in the interstitium may also be responsible for the changes in the lung seen with 'increased permeability' edema. Tissue fibronectin is the principal component of the basement membrane which maintains endothelial cells in proper alignment. A depletion of this substance may be the cause of a capillary leak rather than direct endothelial cell injury. Recent evidence also indicates that depolymerization of the main structural component of the interstitial matrix, namely long hyaluronic acid chains, will also cause edema

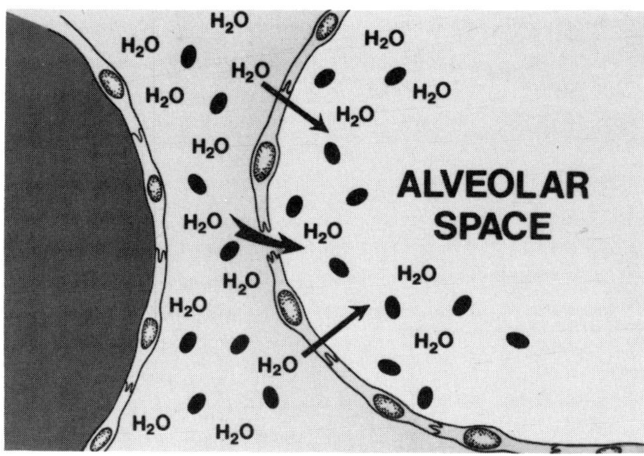

Fig. 4   *The loss of alveolar membrane integrity, leading to alveolar flooding, is usually the result of increased interstitial pressure produced by the accumulation of excess fluid and protein (●). Disruptions of the membrane at cell junctions allow fluid and protein to enter rapidly.*

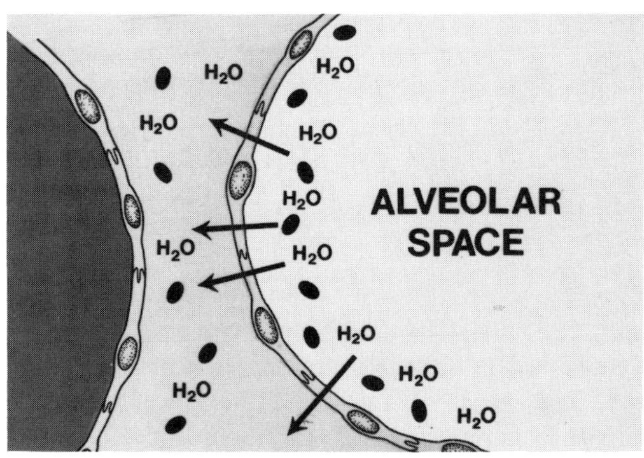

Fig. 5   *When interstitial pressure decreases again, the alveolar cells reoppose each other, restoring normal membrane integrity. Water can cross back into the interstitium, but protein remains trapped, resulting in formation of hyaline membrane and a secondary inflammatory response. This mechanism probably explains why respiratory failure after permeability edema is less easily reversible than that seen after high pressure edema, as more residual alveolar protein is present in the former.*

formation. Oxygen radicals have been found to produce such a depolymerization, and also edema formation, in experimental studies (Cochrane et al 1970; Lee et al 1981). Current information overwhelmingly supports the premise that fluid and protein flux and edema are not simply dependent on the integrity of the microvascular membrane.

3.3. Alveolar membrane

The alveolar membrane is made up of elongated epithelial cells separating the interstitial space and the alveolar gas space. This membrane is also impaired in respiratory failure, which results in alveolar edema and alveolar collapse. The alveolar membrane is much more restrictive to solute flow, allowing essentially no protein to cross under normal circumstances, with the intercellular junction being less than 2.0 nm apart.

Loss of alveolar integrity is usually the result of an expanding interstitial space caused by microvascular injury that increases interstitial pressure and disrupts or pulls apart alveolar cell junctions (Figs. 4–5). The mechanism by which the tight epithelial junctions suddenly give way remains undetermined. This alveolar cell injury is usually readily reversible, with reapposition of cell junctions when the volume and pressure of the interstitial space are decreased. However, the presence of intraalveolar fluid and, in particular, protein, is less readily reversible. Once integrity is regained, intraalveolar water can be reabsorbed, but protein appears to be trapped and can be removed only by macrophages or via transit up the airways. The hyaline membranes characteristic of ARDS are the result of inspissated intraalveolar protein. This explains why permeability edema is more difficult to reverse than high pressure edema. The continued presence of intraalveolar fluid and protein results in the inactivation of surfactant, which produces a progressive alveolar collapse. Direct epithelial cell damage such as can occur with oxygen toxicity or an inflammatory reaction will also lead to alveolar collapse until surfactant is again generated (Clements 1967).

4. GENERAL PHYSIOLOGY OF RESPIRATION

4.1. Ventilation

Maintenance of adequate alveolar ventilation depends upon the status of lung volume. Actually, there are a number of defined lung volumes and capacities as shown in Figure 6 (Comroe et al 1968). It is important to have a working knowledge of these definitions for ventilator management, including indications for placement on and weaning from a ventilator (Table 1; Fig. 6).

Tidal volume is the volume of gas inspired during each respiratory cycle. This is about 4–5 ml/kg of body weight during normal spontaneous breathing.

Total lung capacity is the amount of gas contained in the lung at the end of maximal inspiration. *Vital capacity* is the maximum volume of gas which can be exhaled from the lung after a maximal inhalation. *Functional residual capacity* (FRC) is the volume of gas remaining in the lungs at the resting expiratory level. Normal values for a healthy young adult are shown in Table 1. *Closing volume* is the term used to describe the lung volume at which closure of small airways begins to occur as a result of the elastic recoil properties of the lung. Closing volume is considerably less than FRC under normal circumstances, and therefore small airways remain open during expiration. *Residual volume* is the amount of gas remaining in the lungs at the end of maximal expiration. *Dead space* is the portion of ventilation not contributing to gas exchange. Anatomic dead space includes the trachea and large bronchi. Physiologic dead space includes alveoli that are ventilated but not *perfused*. Normally, the ratio of dead space to total ventilation (V_D/V_T) is about 0.3.

A very important point to recognize is that the airways have a continual tendency to collapse. This is the result of (a) the elastic fibers in the lung that are constantly stretched and attempt to shorten and (b) the surface tension generated by the fluid lining the alveoli as intermolecular attraction continually tries to reduce alveolar volume. This latter effect is reduced by the agent surfactant which normally lines the alveolar surface. This substance, which is made by the alveolar type II cells, is extremely important in preventing alveolar collapse during expiration by decreasing alveolar wall surface tension. In order to maintain airway patency, a pressure difference between the airway and the pleural space has to be generated. It is also important to recognize that the amount of work required to reopen a collapsed airway is greater than the reduction in workload when a collapse takes place.

The expansibility or ease of expansion of the lung is called *compliance*. This is expressed as volume increase per unit increase in transpulmonary pressure and is calculated by dividing the tidal volume by the inflation pressure necessary to achieve that volume. Compliance is determined by both

194

TABLE 1   *Indices of ventilation and perfusion*

	Normal young adult (55–75 kg)	Acute respiratory failure
Vital capacity (ml/kg)	60–75	10
FRC (ml/kg)	30–40	60% of predicted value
Dead Space (V_D/V_T)	0.3	0.6
Compliance (1/cmH$_2$O)	0.1	0.02
Inspiratory force (cmH$_2$O)	$-(80–110)$	-20
Shunt $\dot{Q}_s/\dot{Q}_t$	5%	15%
Pa$_{O_2}$ (room air) mmHg	90–95	60
Pa$_{CO_2}$ mmHg	40–42	60
pH	7.40	7.25
Respiratory Rate	12–18	30

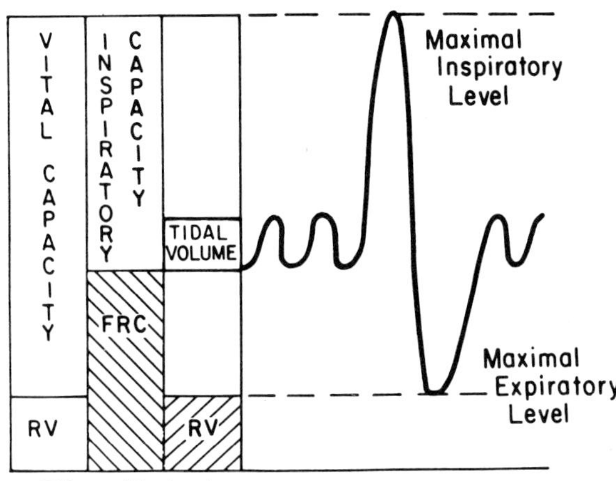

RV – residual volume
FRC– functional residual capacity

Fig. 6   *The interrelationship between ventilatory parameters. The measured values of these parameters can be used to diagnose and manage respiratory failure. RV = residual volume; FRC = functional residual capacity.*

the expansibility of the lung parenchyma itself and of the chest wall, both having to expand readily in order to achieve adequate ventilation. A normal compliance for a young adult is about 0.1 l/cmH$_2$O.

4.2. Perfusion

Despite adequate ventilation, the integrity of the pulmonary circulation is essential in order for adequate gas exchange to occur at the alveolar membrane. Function of the pulmonary circulation can be described in terms of (a) blood flow, (b) blood volume, (c) vascular pressure permeability, and (d) resistance.

4.2.1. Flow

Gas exchange depends on adequate blood flow to areas of alveolar ventilation. Pulmonary blood flow is equal to cardiac output except for some forms of congenital heart disease of arteriovenous shunts. Although oxygen uptake is of major importance, the most important function of the cardiovascular system is adequate delivery of oxygen to the tissues. This is described in the formula: O$_2$ delivery = saturation $\times$ hemoglobin $\times$ 1.39 $\times$ cardiac output $\times$ 10, where 1.39 = ml O$_2$/g of hemoglobin. Note the importance of the cardiac output and hemoglobin concentration.

4.2.2. Volume

Pulmonary capillary blood volume is dependent on the status of total blood volume, pulmonary pressures and resistance. The status of capillary blood volume determines the vascular surface area available for gas exchange.

4.2.3. Vascular pressure and permeability

This was discussed in Section 3.

4.2.4. Resistance

Regulation of pulmonary resistance is determined by the autonomic nervous system and by a number of vasoactive mediators. Changes in resistance will change Pmv. This, in turn, will change blood flow as flow equals the driving

196

pressure divided by the resistance. Changes in resistance can be an important defense mechanism, one of the most significant being the local pulmonary arterial vasoconstriction which occurs in response to a local hypoxia in a poorly ventilated portion of the lung. This results in a shift of blood flow away from areas of poor ventilation. However, inappropriate changes in resistance in the pathologic state can also be quite deleterious.

5.　VENTILATION—PERFUSION MATCHING

In order for effective gas exchange to occur, areas of the lung being perfused must be ventilated and vice versa. Even in the normal lung, there is some ventilation–perfusion mismatch in that there is relatively more ventilation to the upper lobes of the lung and relatively more perfusion to the lower lobes. This mismatch can be magnified by ventilatory abnormalities such as bronchoconstriction, alveolar collapse or flooding (pulmonary edema) or perfusion abnormalities such as vascular occlusion of hypoperfusion from hypovolemia. Uneven blood flow can occur as the result of a number of factors. This includes vasoconstriction secondary to sympathetic discharge. Since innervation is uneven, an uneven pattern of constriction can occur. A number of vasoactive substances such as the prostaglandins can cause variable constriction and dilatation, again leading to uneven blood flow. When an area of the lung with no ventilation is perfused, no oxygenation occurs. This is known as a *shunt* and the amount of total pulmonary blood flow which perfuses nonventilated areas is known as the *shunt fraction* ($\dot{Q}_s/\dot{Q}_t$). In normal lungs, $\dot{Q}_s/\dot{Q}_t$ is less than 5%. Shunt in excess of 15% will lead to a significant impairment in oxygenation. A true shunt does not respond to oxygen administration because the excess oxygen cannot reach the blood through a nonventilated area. When an area of the lung is ventilated but poorly perfused, an increase in physiologic dead space V_D/V_T is seen (Comroe et al 1968). Most ventilation–perfusion mismatch is not the result of a total lack of either to an area, and therefore, some response to exogenous oxygen administration is usually seen.

6.　DEFINITION OF ACUTE RESPIRATORY FAILURE

Respiratory failure is defined as the clinical state in which pulmonary gas exchange and oxygen delivery are no longer adequate to support body function. There are certain arbitrary measured criteria which can be used to make this diagnosis. These are shown in Table 1. Added to these criteria should be an excessive increase in the work of breathing. Labored respirations reflect increased work and are an indication that fatigue and respiratory failure are imminent. These criteria do not all have to be present before a diagnosis of respiratory failure can be made, but are used more as guides. Once the diagnosis of acute respiratory failure is made, further support of ventilation and/or perfusion is necessary. This usually means endotracheal intubation and mechanical ventilation with supplementary oxygen.

As can be seen by the values in Table 1, there is a tremendous margin of safety between normal lung function and the criteria for respiratory failure. Therefore, a significant degree of lung damage is required before the symptoms occur. During the progression from the normal lung to respiratory failure, an entire spectrum of disease states is often seen. Early measures at prevention or treatment can often reverse the disease process.

7. CHARACTERIZATION OF THE ADULT RESPIRATORY DISTRESS SYNDROME

7.1. Early phase

Adult respiratory distress syndrome (ARDS) is the name given to the indirect lung injury often seen after sepsis (Clowes et al 1974; Demling 1981; Lewis et al 1977). Tachypnea and dyspnea are early signs, progressing to bronchoconstriction, increased work of breathing, ventilation–perfusion mismatch, increased pulmonary shunting and dead space, and eventual hypoxia. Early pathologic findings consist of interstitial edema, focal hemorrhage, and atelectasis, with pulmonary microvascular congestion. This progresses to intraalveolar edema and hemorrhage with severe congestion and atelectasis. Hyaline membranes are seen in the alveoli (Petty et al 1979).

The early stages of respiratory insufficiency can be explained in part by a loss of microvascular membrane integrity, either by pressure or permeability, progressing to a loss of alveolar membrane integrity (Anderson et al 1979; Brigham et al 1979; Fein et al 1979; Hinshaw et al 1974; Pietra et al 1981). The pathophysiology of most sepsis injury appears to result from a combination of (a) membrane alterations with congestion and hemorrhage caused by increased pressure and edema due to changes in permeability, and (b) airway

198

constriction and alveolar collapse. Later changes which may occur in the interstitium as a result of protein depletion or damage to the matrix, remain poorly defined at this point in time (McCraig and Parratt 1980).

The significance of airway abnormalities in ARDS has up till now been underestimated. Although extravascular lung water has been reported to be increased in both man and animals during this injury, there is not an absolute correlation between the amount of lung water and the degree of physiologic shunting (Brigham et al 1983; Krausz et al 1977; Lava et al 1982). Constriction of small airways and the loss of the normally protective hypoxic pulmonary vasoconstriction are two processes which are known to occur during sepsis.

The early circulatory and ventilatory lung changes are, for the most part, completely reversible and most patients survive the initial pulmonary insult if sepsis is controlled. When the initial edema and hemorrhage begin to clear, the alveolar hyaline membranes become more prominent. After 4 or 5 days, inflammatory changes are present in response to the inspissated protein. Some areas of destruction are seen resulting from the inflammation or superimposed infection. Epithelial regeneration then occurs. The regenerating cells are predominantly alveolar type II cells, the cells responsible for surfactant (Clements 1967; Robertson 1978). At this stage, between 1 and 2 weeks, most cases of ARDS either rapidly resolve or result in mortality, often as a result of sepsis-related multisystem failure. If early resolution occurs, it is usually complete since there is no irreversible airway injury as is seen with acid aspiration or smoke inhalation. If ARDS persists, a late phase is entered, which is characterized by the presence of acute changes superimposed on a progressive pulmonary fibrosis (Snow et al 1982; Zapol et al 1977, 1979).

7.2. Late phase (pulmonary fibrosis)

When respiratory failure persists, fibroblast proliferation becomes prominent and results in interstitial and intraalveolar fibrosis. Areas of the lung may become relatively acellular, being replaced by fibrous tissue. It has been generally assumed that the fibrosis is associated with an increased collagen concentration. However, there have been conflicting reports as to whether collagen is actually increased in ARDS. Some studies have reported the actual collagen content to be normal. Recently, Zapol et al (1979) have reevaluated this problem. They found that the concentration of collagen in the lung parenchyma of patients with prolonged ARDS was clearly increased. However, nonparenchymal contaminants brought into the lungs from acute inflammation diluted the collagen in the tissue samples taken for chemical analysis. When corrections were made for contaminants such as hemoglobin, collagen levels were significantly elevated. The reason for the increased levels of collagen remains undetermined. Net collagen deposition may be a result of increased rate of synthesis, decreased rate of degradation, or both. The newly regenerated alveolar cell is extremely sensitive to a second injury (Demling 1981). These cells respond to second injury, whether it is due to a superimposed sepsis or to excess oxygen, by the deposition of collagen. Electron microscopic studies have shown that these cells contain an extensive endoplasmic reticulum, suggesting a high rate of protein synthesis. It is very likely that these cells are responsible for excess collagen deposition. Excess collagen and impaired formation of surfactant result in the characteristic lung stiffness.

Although it is a difficult management problem to maintain satisfactory gas exchange, it is of extreme importance to recognize that this fibrotic process is also reversible. Long-term studies have shown that in spite of the potential for permanent respiratory impairment, if the patient ultimately recovers, lung function will return toward normal. This has been validated in patients with multisystem trauma and in burn patients. The problem occurs when the fibrosis is so severe that gas exchange cannot be maintained or the patient cannot be weaned from the ventilator.

8. PATHOPHYSIOLOGY OF SEPSIS-INDUCED ADULT RESPIRATORY DISTRESS SYNDROME

There are three characteristics of sepsis-induced ARDS in man. The pulmonary circulatory characteristics include pulmonary artery hypertension and increased pulmonary capillary permeability (Finley et al 1975; Sibbald and Driedger 1983). The combined circulatory abnormality, as described in Figure 2, markedly potentiates the increased fluid and protein flux seen with either injury separately. The degree of pulmonary hypertension has been found in man to correlate directly with subsequent mortality (Sibbald et al 1978). Although the majority of the increase in pulmonary vascular resistance appears to be arteriolar, there is an increase in Pmv as well, as reflected by an increase in pulmonary capillary wedge pressure and an increase in protein-poor lung lymph seen in a number of animal models (Fig. 7) (Brigham et al 1979; Demling et al 1981a). When severe (mean pressure exceeding 30 mmHg), the pulmonary hypertension can lead to right ventricular dysfunction resulting in ventricular septal deviation to the left. This, in turn, will lead to a decrease in left ventricular compliance and in cardiac output (Parwitt and Ghignone 1983).

An increase in protein permeability is also characteristic of sepsis or endotoxin injury in most animal species (Brigham et al 1979; Demling et al 1981a, 1981b; Fein et al 1979). There is recent documentation that this occurs in man as well. Several studies directly sampling edema fluid have reported protein contents approaching the plasma value (Anderson et al 1979; Staub

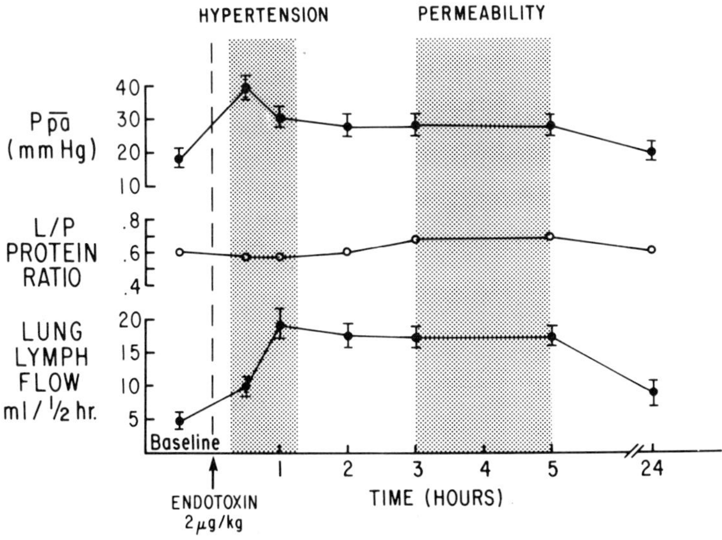

Fig. 7   *The mean ± standard deviation (SD) response of the pulmonary microcirculation to an intravenous infusion of E. coli endotoxin 2 µg/kg for 8 unanesthetized sheep with lung lymph fistulas. A two-phase response is seen. Severe pulmonary artery hypertension occurs, beginning about 30 minutes after endotoxin and lasting for about 1 hour. Lung lymph flow is significantly increased during this period. The lymph-to-plasma (L/P) total protein ratio decreased slightly during this period, indicating some membrane sieving. At about 3 hours, a steady state is reached during which lymph flow is increased, but with protein-rich lymph as the L/P ratio is increased. This indicates an increase in vascular permeability. The steady state lasts 4–6 hours, after which parameters gradually return to baseline (Demling et al 1982).*

1974). Preliminary clinical reports using the in vivo technique of computerized gamma scintigraphy have also indicated a lung capillary leak with sepsis (Sugarman et al 1981).

The third characteristic of septic injury is a ventilatory abnormality (Demling et al 1981a; Kuchelt et al 1980). The increased shunt fraction which is seen is not totally the result of increased lung water. A number of studies have now verified that there is not a direct correlation between shunt fraction and measured water content in the sepsis-injured lung (Brigham et al 1983; Lava et al 1982). Small airways constriction and decreased lung compliance certainly occur in animals after endotoxin. A portion of the sepsis-induced ventilation–perfusion abnormality responsible for hypoxia is felt to be the result of an impairment of the protective mechanism of hypoxic pulmonary vasoconstriction (Lava et al 1982).

A number of etiologic factors are known to participate in indirect lung injury (Clowes et al 1970). Three interrelating elements known to contribute to injury (Fig. 8) are: (a) products of arachidonic acid metabolism, (b) white cells and their byproducts, specifically oxygen radicals, and (c) platelets. These elements clearly interact in producing injury. There is also increasing interest in the relationship between tissue fibronectin, the cement holding endothelial cells together, and lung injury.

Each one of the major elements will be discussed since pharmacological manipulation of these elements will shortly be feasible in man.

8.1. Arachidonic acid metabolites

Arachidonic acid, a component of all cell membranes, is released during sepsis or cellular injury through the action of a variety of phospholipase enzymes on the cell membrane. The released arachidonic acid then enters one of at least three different pathways. The most common and best described pathway is via the enzyme cyclooxygenase which results in the formation of prostaglandins. The second pathway which has recently generated considerable enthusiasm is that via the enzyme lipoxygenase resulting in the potent vasoactive substances known as the leukotrienes. Arachidonic acid can also be acted upon directly by free oxygen radicals to produce a potent chemotactic agent (O'Flaherty et al 1979).

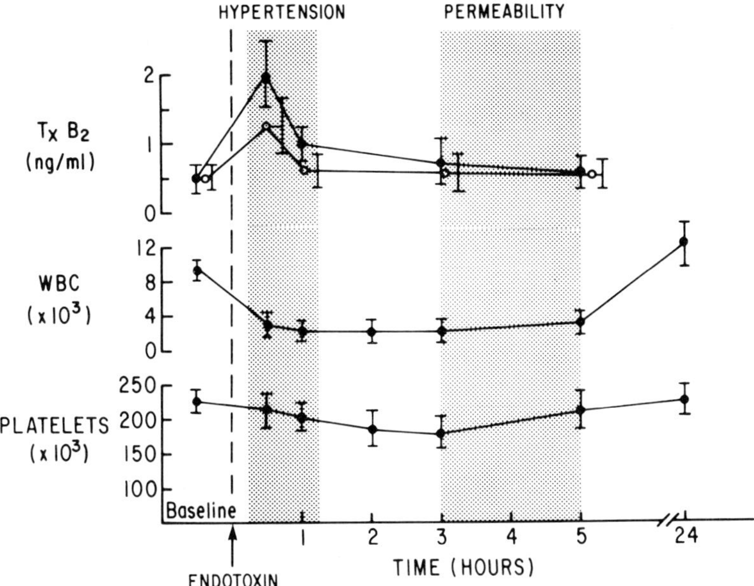

Fig. 8   *The response of blood elements to endotoxin for 8 sheep. TXB$_2$ in lung lymph (closed circles) and in plasma (open circles) is shown. TXB$_2$ levels were significantly increased in the hypertensive phase with lymph values being greater than plasma. Both returned to baseline by the permeability phase. White blood cell count decreased during the hypertensive phase, then remained depressed. Platelets remained relatively stable during the hypertensive phase, decreasing transiently during the early permeability phase. The source of the thromboxane remains undetermined.*

8.2.   Cyclooxygenase pathway products

Prostanoids, or prostaglandins, are vasoactive byproducts of this pathway. The lung is an important organ in the prostanoid system, since it possesses the enzymatic capacity to synthesize all of the prostanoids known presently and is also responsible, in large part, for catabolizing prostaglandins in circulating blood.

Prostanoids are potent compounds capable of significantly altering pulmonary hemodynamics. Any change in the pattern of prostanoid production may be expected to have physiologic significance (Demling et al 1981a, 1982; Flynn and Demling 1982; Moncada and Vane 1979; Winslow et al 1973).

The two most potent prostaglandins are prostacyclin (PGI_2) and thromboxane A_2 (TXA_2). PGI_2 is a powerful inhibitor of platelet aggregation and also acts as a vasodilator. The compound is synthesized by blood vessel walls. TXA_2 is a potent stimulator of platelet and leukocyte aggregation and can also produce severe pulmonary vasoconstriction; it also appears to produce bronchoconstriction (Smith et al 1981a, 1982).

Levels of both these substances are increased after trauma and sepsis (Adams and Traber 1972; Anderson et al 1972; Nerlich et al 1983). Thromboxane is released from platelets and stimulates further platelet aggregation. This substance may well be the active mediator producing pulmonary damage from platelet emboli. Thromboxane is also released from leukocytes and from the lung itself in response to injury. PGI_2 is probably released by the vessel wall in response to injury. When the endothelium is significantly damaged, production of PGI_2 is impaired. This will allow platelets to adhere to the damaged wall and release TXA_2, promoting further platelet aggregation.

The mechanism of action of prostanoids on platelets and endothelial cells is felt to be through the alteration of intracellular cyclic adenosine monophosphate (cAMP) and cyclic guanosine monophosphate (cGMP), with the ratio of AMP to GMP determining the activation state of these cells.

A number of animal studies have now demonstrated that thromboxane release is largely responsible for the early severe pulmonary hypertension and bronchoconstriction seen after endotoxin or bacterial infusion, as inhibition of thromboxane production markedly attenuates this aspect of lung injury (Demling et al 1982). PGI_2 is also found in high concentrations in plasma and in lung fluid following endotoxemia. This is felt to be a protective mechanism in an attempt to negate the TXA_2 effect. Several recent studies in man have now reported an increase in plasma TXA_2 during sepsis-induced ARDS (Parratt et al 1982). Attempts at thromboxane inhibition in man are currently hampered by the lack of a suitable parenteral agent which can be used for human experimental use.

Prostaglandins do not appear to cause the increase in pulmonary capillary permeability which is also characteristic of ARDS. Cyclooxygenase inhibitors have shown no effect on this phase of injury in animal studies. In fact, this injury has actually increased with the use of some cyclooxygenase inhibitors. It has been suggested that the latter effect is due to shunting of arachidonic acid metabolism into the lipoxygenase pathway.

8.3. Lipoxygenase pathway

The end products of this pathway are the leukotrienes (Samuelsson et al 1980). These newly discovered agents, as well as their intermediates, hydroxy and hydroperoxy fatty acids, are potentially deleterious in that they are potent vaso- and bronchoconstrictors. Some of the leukotrienes have also been reported to produce an increase in systemic vascular permeability. The slow reactive substance of anaphylaxis SRS-A is now known to be one of the leukotrienes. The agents are also potent neutrophil chemoattractants and are released in large quantities into inflammatory tissue (Craddock et al 1977; Heflin and Brigham 1981). Neutrophils certainly play a major role in increasing lung vascular permeability. They produce large quantities of leukotrienes which, in turn, attract more granulocytes. These agents are also extremely potent coronary artery constrictors.

The role of the leukotrienes in sepsis lung injury remains largely undefined because of the lack of assay techniques needed to determine their presence in disease. There is evidence that 5-HETE, a product of the neutrophil lipoxygenase system, is present in high concentrations in sheep lung lymph during the increased permeability phase endotoxemia (Ogletree et al 1981). It is also clear that there is a close interaction between the lipoxygenase and the cyclooxygenase pathway and that a significant portion of the injury produced by the infusion of leukotrienes into animals is due to a concomitant release of cyclooxygenase products, especially thromboxane. Lipoxygenase products are also released during inflammation and although they may not have a direct role in a pure endotoxin or sepsis lung injury, they may play a more important part in the accentuated lung injury which appears after soft tissue trauma or a burn, both of which result in an inflammatory focus (Ogletree et al 1981; Samuelsson et al 1980; Smith et al 1981b).

8.4. White cells

There is rapidly accumulating evidence that leukocytes are a major factor in the lung injury from sepsis (Craddock et al 1977; Fletcher and Ramwell 1981; Heflin and Brigham 1981; Hinshaw et al 1967; Kux et al 1972; Milligan et al 1974; O'Flaherty et al 1979). Pulmonary leukostasis is particularly prominent during sepsis and is a characteristic of endotoxemia (Coalson et al 1970; Smith et al 1981a; Zapol et al 1979). When activated, leukocytes release factors such as proteases, lysosomal enzymes, and, of current interest, oxygen radicals which can injure the lung. In a number of animal studies we found that the degree of permeability injury correlated well with the concentration of lung lysosomal enzymes (Demling et al 1980).

Neutrophils appear to play the predominant role in increasing lung vascular permeability in sepsis and other disease states such as oxygen toxicity, as removal of the neutrophils before administration of the insulting agent eliminated the increased permeability in a large number of animal studies (Flick et al 1979; Heflin and Brigham 1981; Wedmare and Williams 1981). Electron micrographic studies have also revealed pulmonary leukocytes in large numbers in patients with ARDS, particularly those with sepsis. Bronchoalveolar lavage in similar patients yielded fluid with increased numbers of polymorphonuclear leukocytes, as well as high levels of neutrophil elastase (Lee et al 1981).

It has now been demonstrated that ARDS in patients on renal dialysis is due to neutrophil activation as a result of blood contact with the dialysis membrane (Craddock et al 1977). It has been stated that activation of the complement pathway may be one of the causes (Hammerschmidt et al 1980; Heidman 1979; Hinshaw et al 1967). The complement system is a group of enzymatic proteins which serve as a defense against foreign elements, most importantly bacteria (McCraig et al 1979). When the complement cascade is initiated by stress situations such as trauma, burns, or sepsis by the so-called 'alternative pathway,' the neutrophil-aggregating complement fragment C5a is released into the intravascular space (Meyer 1973). Activated complement alone is insufficient to produce endothelial damage, but neutrophils are a necessary element: animals rendered neutropenic in the presence of large quantities of C5a had no injury. Corticosteroids in large quantities can inhibit the C5a-induced neutrophil aggregation and subsequent release of toxic substances. There remains considerable controversy in the literature as to just how important the complement cascade is in ARDS. Is complement activation a cause or simply an effect of severe injury and sepsis? Complement factors may produce neutrophil-aggregation as do a number of other agents, but aggregation is not the same as activation.

The presence of neutrophils in the lung by itself does not produce permeability damage. The cells must be activated to release products such as proteases and oxygen radicals, but sequestration in the lung is a necessary element in the injury. A number of chemoattractants are released after trauma or sepsis which may be responsible. These include components of the complement cascade, specifically C5a and most of the leukotrienes. Thromboxane may also have some neutrophil-aggregating properties (Parratt et al 1982). Neutrophil adherence or the stickiness of the cells is also an important property which contributes to the margination of these cells in the microcirculation. Complement fragments have also been reported to increase neutrophil adherence to endothelial cells.

Alveolar macrophages probably play a significant role in ARDS. These cells are known to release potent chemotactic factors. Macrophages are also the cells which are felt to initiate the interstitial inflammatory response.

A number of proteases and lysosomal enzymes are released from activated leukocytes when killing bacteria. Those agents could also alter the endothelial lining and interstitial matrix if released extracellularly, which results in edema formation (Bedrossean et al 1978). However, the release of oxygen radicals, namely O_2^-, (superoxide anion), $OH^\cdot$ (hydroxyl radical) and H_2O_2 (hydrogen peroxide) appear to be the most damaging (Cochrane et al 1970; Sachs et al 1978).

Oxygen radicals are essential for killing of bacteria once they are phagocytized by the neutrophil. These extremely reactive agents can, however, be very destructive when released into tissues (Klebanoff 1980). Oxygen radicals can alter the lipid layer of the cell membrane by increasing peroxide formation. Increased lung lipid peroxidation has been reported in man after a number of insults, including oxygen toxicity and paraquat poisoning. Paraquat is a herbicide which produces a severe acute lung injury as a result of oxygen radical release, eventually leading to severe fibrosis. Lipid peroxidation occurs in the liver after sepsis and may occur in the lung as well. We have recently demonstrated an increase in lipid peroxidation in the sheep lung after endotoxin (Wong et al 1984), which must not be taken as reflecting a cause and effect relationship as the permeability change could be the result of a number of etiologic agents. Oxygen radicals have also been found to enzymatically depolymerize the hyaluronic acid in the lung interstitial space, thereby loosening the matrix and increasing edema formation (Greenfield et al 1974).

Lung edema formation has been described in a number of experimental models where oxygen radicals are liberated. Anti-oxidant agents such as superoxide dismutase and catalase have been reported to eliminate the endothelial injury caused by activated complement and neutrophils, which indicates that the injury is the result of oxygen radical release (Crapo and Tierney 1974). The mechanism of injury, both to the endothelium and the interstitium, remains undefined. Oxygen radicals also cause a rise in thromboxane production, thereby increasing pressure in an injured capillary bed. Radicals act directly on arachidonic acid to produce a potent chemotactic lipid which can cause further neutrophil sequestration.

Recent studies, in which chemiluminescent activity was measured as an indication of release of oxygen radicals, have reported that the circulating leukocytes in patients with ARDS are indeed activated and release oxygen radicals (Zimmermann et al 1983).

8.5. Platelets and the coagulation system

Platelets and the products of coagulation are also felt to be etiologic agents in respiratory failure (Busch et al 1973). Although platelet and fibrin microaggregates have been frequently found in the pulmonary microcirculation after injury, their significance in producing lung injury remains controversial (Busch et al 1971, 1974; Saldeen 1982). In certain hematologic diseases, such as thrombotic thrombocytopenia purpura, massive platelet embolization occurs, resulting in respiratory failure. This indicates that platelets are capable of producing lung damage by themselves. However, platelet-induced injury after sepsis has been more difficult to document. Because of the other events that occur simultaneously after major injury which can also produce lung damage, such as complement activation, lysosomal enzyme release, and prostanoid production, it has been difficult to determine the role of the platelets. The problem is further complicated by the fact that these aggregates are frequently transient and not always evident on post-mortem examination. The major issue, however, is not whether pulmonary microemboli exist, but rather whether platelets or products of coagulation actually injure the microvascular membrane (Giordino et al 1976). Hechtman et al (1978), for example, found no correlation between the pulmonary entrapment of platelets in patients with ARDS and any functional abnormality. Others have found that removing platelets did not result in any protection (Malik et al 1982).

There are two possible mechanisms by which microemboli themselves could produce injury. First, damage could be produced by physical obstruction of the microcirculation leading to a 'high pressure' form of edema in the remaining intact perfused vessels. It has been shown that experimental occlusion of a large portion of the pulmonary vascular bed by emboli or balloon produces a modest increase in permeability in the remaining vessels (Binder et al 1979). However, this requires a degree of vascular occlusion sufficient to increase pulmonary vascular resistance 2–3-fold. It is unlikely, therefore, that simple obstruction is a major source of pulmonary damage in sepsis.

The second and more plausible explanation is that humoral factors are released, either as a direct result of embolic occlusion or from elements in the emboli such as platelets or products of the coagulation cascade, which results in membrane damage (Aasen et al 1978). Malik et al (1982) demonstrated that even inert emboli such as glass beads stimulate the release of a factor that modestly increases lung permeability. However, lung water did not increase, which indicates that injury is minimal.

Platelets have also been found to release a number of factors that can alter lung function. Serotonin is a vasoactive agent found in high concentrations in platelets (Vincent et al 1983). When released, this substance results in bronchoconstriction, increased pulmonary venous pressure, and increased lung water. Capillary pressure is increased because of increased venous resistance, and the result is a 'high pressure' form of edema (Moss et al 1973). Recently, platelet serotonin has been found to accentuate neutrophil adherence and release, thereby augmenting the neutrophil-induced endothelial injury (Boogaerts et al 1982). Histamine is found in platelets in man but not in particularly large quantities. Histamine infusions produce bronchoconstriction but no increase in pulmonary vascular permeability. Lung mast cells are probably the major source of the histamine that is released, rather than platelets. Thromboxane is released by platelets. Although there is good evidence that platelets are involved in ARDS, there is also contrary evidence. Busch et al (1971) reported that removing platelets did not prevent the lung injury produced by microembolization of glass beads, and thrombocytopenia is not a universal finding in experimental or clinical ARDS.

The coagulation products, fibrin and fibrin factors, have been reported to cause lung injury in animals. Malik et al (1982) reported that pulmonary injury produced by glass bead embolization could be prevented by heparin. Injury was prolonged when fibrinolysis was blocked. However, Binder et al (1979), using a similar model, were unable to reproduce this effect. Busch noted that defibrinogenation prevented the lung damage induced by thrombin infusion. Infusion of the fibrin-split product fragment D has been reported to produce severe lung injury in rabbits, but injury in other animals has not been reported. At the present time, the role of coagulation and microembolization in lung injury remains controversial.

8.6. Depression of reticuloendothelial system and fibronectin deficiency

The reticuloendothelial system, in particular the liver, is responsible for clearing some bacteria–platelet–fibrin aggregates or other forms of microemboli that are felt to be important during sepsis. Recently, a marked depression of reticuloendothelial function has been found in septic patients. This failure to clear particles results in an increase in particle embolization of the lung. The major site of reticuloendothelial clearance of blood-borne particulates is the hepatic Kupffer cell. Plasma fibronectin has a high affinity for collagen and fibrin complexes and has been reported to facilitate clearance. The level of fibronectin is severely depressed after sepsis, resulting in an opsonic deficiency and therefore in decreased reticuloendothelial clearance. Patients with sepsis with this deficiency have often demonstrated coexistent pulmonary distress. Saba et al (1979) have reduplicated this deficiency in experimental animals and found that lung injury was exaggerated in septic patients with the reticuloendothelial deficiency. The correlation between depressed fibronectin and pulmonary injury in the traumatized patients has been corroborated by Scovill et al (1979), but a cause and effect relationship remains to be demonstrated (Maunder et al 1983).

Another form of fibronectin is known as 'tissue fibronectin'. This protein is considered to be the cell cement maintaining the basement membrane and the tight endothelial cell junctions. A deficieny or destruction of this protein would result in widening of these junctions and increased vascular permeability. Leukocyte proteases released in the process of phagocytosis have been found to destroy the pulmonary tissue fibronectin and may be another source of pulmonary edema during sepsis (McDonald et al 1979).

The role of both forms of fibronectin in the maintenance of pulmonary function remains somewhat undefined as it is difficult to determine whether fibronectin levels are simply decreased in response to severe sepsis or the decrease actually results in organ dysfunction. It is highly probable that the tissue fibronectin plays a more important role in maintaining lung function than does the plasma form, but there are no data at the present time which indicate that (a) depletion of this form in the lung leads to increased permeability after trauma or sepsis, or (b) infusion of the plasma form can correct a tissue deficit, as the two forms have minor but significant structural differences. As can be seen from this discussion, sepsis-induced ARDS is a complex entity with multiple interrelated causes, whose specific pathophysiology remains to be defined.

9. MYOCARDIAL DYSFUNCTION

Although hemodynamic instability during sepsis is largely the result of volume depletion, some of the alterations in cardiac output which occur in sepsis, particularly late sepsis, appear to be due to direct myocardial dysfunction (Cann et al 1972; Goldfarb et al 1979; McCraig et al 1979; Raffa and Trunkey 1978). These direct changes from sepsis or endotoxin are not well-defined and remain somewhat controversial. In early sepsis, there is an increase in cardiac output and in pulse rate which may be due to increased beta adrenergic stimulation. In the later stages of sepsis, the response of the myocardium to epinephrine seems to be reduced. This may reflect impaired function of simply maximum physiologic stimulation. These changes explain why stroke work cannot return to normal by beta stimulation in low output septic shock. There is now evidence that mitochondrial energy production is impaired with endotoxin (see Volume 3 of this series, Chapter 7).

Direct mycocardial depression from a circulating factor has been said to cause myocardial dysfunction (Greenfield et al 1974). The pancreas has been reported as one source of a sepsis-induced substance. It has also been suggested that endotoxin itself is a cardiac depressant, or β-endorphins released during sepsis in excess amounts (Reynolds et al 1980).

The presence of a circulating depressant substance has not been universally demonstrated or widely accepted, and remains speculative, as restoration of an adequate circulating blood volume appears to correct many of the abnormalities.

Right ventricular encroachment on the left ventricle, due to septal deviation, is now recognized as a significant cause of decreased cardiac output in the critically ill patient (Calvin et al 1981; Parratt et al 1982). A number of recent studies in man and animals have clearly demonstrated that sepsis-induced pulmonary artery hypertension can lead to right heart strain and, in turn, a decrease in left ventricular performance. This appears to be due to a shift of the ventricular septum to the left, thereby decreasing left ventricular compliance. This does not appear to occur in animals until mean pulmonary artery pressure reaches 30 mmHg. High levels of positive end expiratory pressure can also cause the same process (Calvin et al 1981).

10. ABNORMALITIES IN THE SYSTEMIC CIRCULATION

The hemodynamic changes during sepsis have been well described in man and animals (Blain et al 1970; Hess et al 1981; Hinshaw et al 1974; Wilson 1972). The initial response of the septic patient or animal, given adequate fluids, is an increase in capillary perfusion to skin and kidney. This initial 'hyperdynamic state' is characterized by a decrease in systemic vascular resistance and an increase in cardiac output (Goldfarb et al 1979; Krausz et al 1977). This appears to be caused by inflammatory mediators which produce a vasodilatation. A number of vasoactive agents are released from the area of sepsis. These include bradykinin, vasodilatory prostaglandins and histamine. All of these agents significantly potentiate the actions of the other substances. Bradykinin is a particularly potent vasodilator which probably causes much of the flushing seen in sepsis. Although cardiac output is increased, oxygen delivery to the tissues is often decreased in sepsis, as evidenced by a decrease in oxygen consumption (Duff et al 1969). This appears to be caused by the opening of low resistance arteriovenous channels bypassing the capillary bed. The mechanism of the shunting of blood is not well understood, but also appears to be the result of inflammatory mediators.

As the sepsis progresses, the increasing vasodilatation now results in a relative hypovolemic state. An increase in net transvascular fluid flux has also been reported to occur, causing further plasma loss and leading to interstitial edema. Three mechanisms can be in part responsible for this continuing fluid loss. The first is an increase in systemic vascular permeability. This is particularly evident in an area of inflammation. Proteases and other lysosomal enzymes released from leukocytes can lead to increased permeability as can other agents, such as histamine and bradykinin. Oxygen radicals released from activated leukocytes can result in damage to the systemic circulation in a similar fashion as reported for the lung (Weismann et al 1980). Interestingly, we have not identified a systemic capillary leak after endotoxin during the period when a lung leak was present. This process may be only an end stage effect due to prolonged acidoses and tissue ischemia (Kramer et al 1982a).

The second mechanism would be an increase in capillary hydrostatic pressure. Agents such as serotonin and the leukotrienes are potent venoconstrictors which increase venous resistance and, in turn, capillary hydrostatic pressure. Pressure and permeability in combination of course substantially potentiate each other.

The third mechanism is through an alteration in the interstitium loosening the matrix, thereby increasing the ease of edema formation (Comper and Laurent 1978; Greenwald and Moy 1980; Kramer et al 1981; Lee et al 1981). We have recently demonstrated that hypoproteinemia results in significant edema accumulation in soft tissues such as skin and muscle (Kramer et al 1982a). This effect is much more prominent in these tissues than in the lung. Interstitial protein depletion appears to be the initiating factor. It seems that the increased fluid flux is caused by a hydration and loosening of the interstitial gel and is totally reversible with protein replacement. Hypoproteinemia is, of course, a common finding in septic patients, particularly with fluid resuscitation.

As the fluid loss progresses, a hypodynamic state results, characterized by hypoperfusion, oliguria, and decreased cardiac output. The release of catecholamines in response to the hypovolemic state now produces an increase in peripheral vascular resistance, a decrease in cardiac output, and symptoms similar to that seen with any severely hypovolemic state, including hypotension and oliguria. This would be defined as septic shock, which is the end state of the septic process. Activation of the clotting cascade (Hardaway 1973) and complement cascades (Heidman 1979), which occurred early in sepsis, will potentiate the progressive tissue ischemia.

11. GENERAL PRINCIPLES OF PREVENTION AND TREATMENT OF SEPSIS-INDUCED CARDIOPULMONARY DYSFUNCTION

11.1. Surgical treatment

Most cases of sepsis-induced cardiopulmonary failure are the result of the presence of devitalized tissue, infected or noninfected. Both are sources of mediators of inflammation which, in turn, can lead particularly to acute lung injury. The early debridement of dead tissue and the early drainage or excision of infected tissue will do more to diminish the severity of sepsis and incidence of sepsis-related mortality than any new pharmacological agent. This should be carried out as soon as the patient is stabilized with fluids and antibiotics are initiated (Weinstein and Skillman 1981).

11.2. Monitoring techniques

There are a number of parameters which can be monitored to allow for the early detection of ARDS. *Chest X-rays* by and large are not an early detector of injury as abnormalities in the X-ray frequently post-date clinical symptomatology by 24 hours. It has also been shown that X-rays do not accurately quantitate changes in lung water as detected by the double indicator dilution techniques (Baudendistel et al 1982). Chest X-rays can frequently assist in the determination of the cause of changes in lung function, whether they be due to focal ventilatory abnormalities or diffuse circulatory abnormalities such as edema. *Blood gases* are probably the least invasive, yet reliable parameters for detecting early lung injury. This coupled with a few simple *pulmonary function tests*, such as respiratory rate, tidal volume and maximum inspiratory force, may allow one to determine both the presence and the cause of early lung failure. Normal values are presented in Table 1. A decrease in arterial oxygen tension, of course, indicates either a diffusion abnormality of the alveolar capillary interface as produced by edema, or an increase in 'shunt fraction' due to airway closure. A decrease in *tidal volume*, either measured or determined by observation of chest wall motion, will help determine whether hypoventilation is present. This can be caused by pain or excess sedation or by generalized airway disease. The latter is usually characterized by some wheezing on physical exam. Changes in *arterial carbon dioxide* content can also be used to detect early injury. A decrease in Pa_{CO_2} is characteristic of the early phase of injury. The hyperventilation seen in sepsis appears to be due to a direct effect of endotoxin on the brain. An increase in Pa_{CO_2} reflects an increase in dead space ventilation as can occur with vascular occlusive disease or a greater ventilation–perfusion mismatch from vasoconstriction or edema. An alteration in blood gases is frequently not the first symptom of lung injury but reflects an exhaustion of normal compensatory mechanisms such as increasing work of breathing to move more air in order to maintain reasonable gas exchange in a compromised lung. Therefore, a patient who demonstrates increasing work with slowly deteriorating gas exchange may be a candidate for endotracheal intubation and mechanical ventilation prior to the inevitable onset of measurable hypoxia. If the patient is already intubated, serial calculations of shunt fraction and lung compliance can be made and early abnormalities detected.

Transcutaneous measurement of Pa_{O_2} and Pa_{CO_2} is a noninvasive method whereby blood gases can be continuously monitored. The technique involves the heating of the skin beneath a small surface electrode and the measurement of blood gases in the underlying microcirculation. This technique is used extensively in neonatal units as transcutaneously measured values of O_2 and CO_2 have proved to be very accurate indicators of the values in arterial blood since infants have very thin skin. Studies in adults have produced variable results as to the accuracy of the method. It is clear that blood volume is also important in interpreting the data. Current problems with the method can be solved by better technology. Transcutaneous blood gas analysis may prove to be very valuable in the near future (Tremper and Shoemaker 1981).

Maintenance of adequate *blood flow* is essential for oxygen exchange and delivery. This requires the maintenance of cardiac output. Good arterial blood gases mean little if the oxygen is not being delivered to the tissues. In critically ill patients, cardiac output can be inadequate, even in the presence of an increase in central venous pressure and peripheral edema. If the status of perfusion cannot be determined clinically, *cardiac output* should be measured, usually by the thermodilution technique with a Swan-Ganz catheter. *Oxygen consumption* is a good monitor of the adequacy of tissue perfusion and, in turn, of cell metabolism. The difference between arterial and mixed venous oxygen content can be used to monitor oxygen consumption. A normal Ca_{O_2}–Cv_{O_2} difference is about 4 volumes. An increase in the value is frequently seen during the hypovolemic state as transit time through the microcirculation is increased. Frequently, during the hyperdynamic phase of sepsis, the difference is decreased. This is felt to be due to the bypassing of the microvascular bed of shock, which, in turn, can lead to tissue oxygen debt.

Measurements of *pulmonary vascular pressures* and cardiac output require a relatively invasive technique, whereby an indwelling line in the pulmonary circulation and heart is required. This technique is not without major hazards. Septicemia from catheter infection and massive lung hemorrhage from rupture of pulmonary vessels, as well as arrhythmias, occur not uncommonly. The potential for infectious complications of this technique can be diminished if lines are changed every 72 hours.

Hemorrhage and infarction are usually caused by maintaining the catheter in the wedge position for more than a few minutes necessary per measurement. This occurs when the catheter migrates farther into the lung or the balloon is not totally deflated. Cardiac arrhythmias usually occur at the initial placement of the line, rather than as a later event. Adequate monitoring is, of course, necessary during placement.

Detection and quantitation of pulmonary artery hypertension is an advantage in that this characteristic is frequently an early sign of sepsis-induced ARDS. The measurement of *pulmonary artery wedge pressure* is also important in controlling volume resuscitation to avoid accentuating edema formation.

When increased permeability is present, the wedge pressure should be maintained at the lowest possible value needed to maintain an adequate cardiac output.

Pulmonary wedge pressure has always been considered to reflect left ventricular end diastolic volume, LVEDV, which in turn determines ventricular function through Starling's law of the heart. Several recent large clinical series have demonstrated that in a significant number of critically ill patients, wedge pressure does not accurately reflect LVEDV, particularly when there is severe pulmonary hypertension (Martyn et al 1981). This is caused by the deviation of the ventricular septum to the left and, consequently, a decrease in left ventricular compliance. Therefore, a much higher wedge pressure may be necessary in these patients to maintain an adequate cardiac output. However, if ARDS is present as well, further edema will result. Treating the patient with both cardiac and pulmonary disease is truly a 'high wire balancing act' where the maximum benefit for both organ systems must be maintained simultaneously (Hess et al 1981; Siegel et al 1967; Wilson 1972). If one organ function has to be temporarily sacrificed, it should be the lung, as recovery from lung edema is much easier than recovery from prolonged shock.

An increase in *extravascular lung water (EVLW)* is seen in a number of ARDS syndromes, particularly with sepsis (Holcroft 1977; Todd et al 1978; Tranbaugh et al 1982). Several techniques are used clinically to detect changes in lung water in the hope that this is an early and reliable measure of deteriorating lung function. A rebreathing technique with dimethyl ether and helium has been used to measure EVLW; experimental studies have shown that it correlates well with post-mortem lung water. This technique has been used in several clinical studies (Petrini et al 1978).

Thermal dilution-green dye double indicator dilution technique is another method (Lewis et al 1979). Cold, the thermal indicator measures intra- and extravascular water while the green dye only measures the intravascular space. Subtraction of the two areas of distribution leads to a computation of EVLW. This technique also appears to correlate with gravimetrically measured lung water and has been used in a number of clinical studies. Several other techniques for the measurement of EVLW in man are also available.

The major criticism in regard to the accuracy of all these methods is that water in nonperfused areas may not be detected, for example, in areas obstructed with emboli. Of more concern, however, is the limited clinical usefulness of the value measured. There is no question that as an experimental tool for determining, for example, the effect of various fluids on EVLW, these techniques are excellent (Lewis et al 1979), but there does not appear to be a good correlation between hypoxia, shunt fraction, and EVLW (Lava et al 1982). This is logical as abnormalities in ventilation not caused by an increase in EVLW play a significant role in ARDS.

Another more useful technique for monitoring lung injury is the measurement of changes in lung protein permeability by the use of radioactive tracers and a portable gamma camera or by the use of a radionuclide scanning technique called scintigraphy. These methods, which are undergoing clinical trials, have the advantage of distinguishing between pulmonary edema due to changes in pressure and due to changes in protein permeability (Tate et al 1981; Tatum et al 1982; Thoren et al 1978).

11.3. Maintenance of ventilation

Ventilation needs to be maintained by first maintaining an adequate minute ventilation (tidal volume × respiratory rate) to provide for adequate gas exchange. Shallow breathing, even if sufficient to maintain gas exchange, will eventually lead to alveolar collapse unless an occasional hyperinflation is produced. This can be prevented by avoidance of excessive sedation while still providing adequate pain medication to decrease chest wall splinting. Adequate nutritional support is also necessary to provide energy for chest wall and diaphragm muscle activity.

Airway patency, both large and small, needs to be maintained. Large airway closure is frequently the result of large mucous plugs or foreign bodies while small airway closure occurs as the result of (a) airway edema, (b) decrease in function residual capacity (FRC) below the closing volume of small airways, (c) extrinsic compression from interstitial edema, and (d) decrease in the clearance of mucous. Probably the most common cause is a decrease in FRC which occurs as a result of hypoventilation from pain, trauma, or because of mediator-induced bronchoconstriction. When FRC decreases, elastic forces attempting to close the airways can exceed the tethering forces produced by lung expansion, maintaining airway patency. This results in small airway collapse. The volume of the lung at which this occurs, as previously mentioned, is the closing volume.

If vital capacity and FRC cannot be adequately maintained, as reflected by deterioration in blood gases and an increase in shunt, or an increase in the work of breathing, endotracheal intubation and mechanical ventilation is indicated. It is not necessary to wait until the patient meets all the criteria of respiratory failure before this is accomplished, as early mechanical ventilation can prevent further airway closure and an increase in injury. As previously mentioned, it is much easier to keep airways open then to reopen closed airways. The ventilator is usually adjusted to deliver a tidal volume of 10–15 ml/kg body weight.

11.3.1. *Positive end expiratory pressure (PEEP)*

The use of PEEP in patients with respiratory failure results in an increase in FRC and Pa_{O_2} by re-expanding and maintaining expanded, poorly ventilated areas of the lung. This will decrease the shunt fraction $\dot{Q}_s/\dot{Q}_t$. PEEP also improves gas exchange in pulmonary edema where alveoli are filled with fluid (Fig. 9). This is due to the resultant increase in alveolar volume, rather than a decrease in lung water. The positive pressures cause the intraalveolar fluid to be spread in a layer along the alveolar walls thin enough to allow diffusion of gases to and from the capillary. Levels of PEEP of between 5–10 cmH_2O are usually sufficient; however, occasionally higher levels are required. PEEP has to be used carefully as a number of problems can result with its use. First, cardiac output can decrease as a result of a decrease in venous return due to the PEEP-induced increase in positive intrathoracic pressure. If Pa_{O_2} improves but cardiac output decreases, the net effect is a decrease in oxygen delivery to the tissues. Secondly, PEEP can also result in over-expansion of more compliant areas of the lung, leading to an increase in pulmonary vascular resistance and decreased blood flow in these areas. Of particular concern is the result of sudden discontinuation of PEEP in ARDS patients for purposes of suctioning, because it may take up to 5 minutes after reconnection of the ventilator and PEEP for lung function to return to presuction levels (Bodai and Blaisdell 1981). This again emphasizes the difficulty of reopening airways once closed.

Barotrauma or injury to the lung as a result of positive pressure ventilation is also a real entity. Recent reports indicate the incidence of pneumothorax

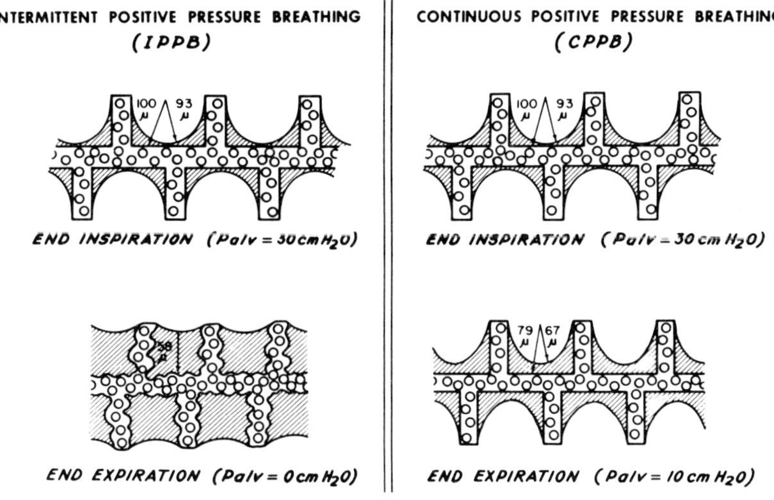

Fig. 9　*Scheme demonstrating probable basis for beneficial effect of positive pressure breathing in improving arterial blood oxygenation in pulmonary edema. In all diagrams, each alveolus contains the same volume of fluid. At $P_{A_V} = 30$ cmH₂O, alveoli fill with air and edema fluid layer is reduced to 7 μm. At end expiration, alveoli collapse to low volume with a 58 μm thick fluid barrier, if $P_{A_V} = 0$ (intermittent positive pressure), but remains partly gas-filled with only 12 μm thick fluid barrier if $P_{A_V} = 10$ cmH₂O (continuous positive pressure). There is no evidence that either form of positive pressure shifts edema fluid out of lung. Reprinted from Staub (1974), with permission.*

and resulting bronchopleural fistulas to be in the order of 5–25% of all ventilated patients. Therefore, mechanical ventilation, particularly with PEEP, may be very beneficial but is certainly not without risk.

An alternative to *high pressure ventilation* was reported by Sjorstrand (1980) and by many other groups. It has been demonstrated that mechanical ventilation with a volume of 50–200 ml at a rate of 80–100 breaths per minute can support gas exchange but with much lower transpulmonary pressures and, therefore, elimination of barotrauma. Ventilation can be maintained by administering jets of air through 14–18 gauge cannula, as opposed to the need for an endotracheal tube with its secondary tracheal irritation. Clinical trials have shown that FRC can be maintained without the need for PEEP. Another advantage of the system is the ability to suction an individual without disconnecting him from the ventilator as the cannula is small enough to introduce a suction catheter alongside.

As one might expect, the machinery necessary to deliver the *high fre-*

quency jet ventilation is complex, and considerable research is still needed to determine which patients would benefit from its use. It appears clear at this time that one group of patients will certainly benefit, namely those patients with large broncho-pleural-cutaneous fistulas or tracheo-esophageal fistulas, which do not tolerate positive pressure ventilation because of the accentuation of air leak.

11.3.2. Nutrition

The patient with ARDS can develop weakness of the chest wall respiratory muscles of such magnitude that weaning from a ventilator becomes extremely difficult. Muscle weakness occurs for two reasons: first, as the result of the accelerated catabolism seen in the critically ill patient, and second because of disuse atrophy after placement on controlled mechanical ventilation for any extended period of time. Adequate nutritional support will help control the catabolism, and the use of intermittent mandatory ventilation, instead of total ventilatory support, will exercise the chest wall muscles and help prevent the muscle atrophy.

An *increase in carbondioxide production* will require an increase in ventilation to avoid hypercarbea (Kinney et al 1980). This can lead to an increase in the work of breathing beyond that which the compromised patient can tolerate without mechanical ventilatory support. Although the administration of adequate calories is essential, excessive amounts of calories administered as carbohydrates lead to the formation of fat and the liberation of excess carbondioxide. Respiratory distress, precipitated by overzealous carbohydrate administration beyond that which the patient can use for energy, is seen in increasing frequency because of the increased usage of nutritional support. This process is readily reversed by decreasing the carbohydrate load.

11.3.3. Oxygen administration

Use of high concentrations of oxygen is frequently necessary in the management of respiratory insufficiency. Prolonged exposure to high concentrations of oxygen results in lung damage and further respiratory failure.

Because of these hazards, it is important to decrease the fractional concentration of oxygen (FiO_2) below 50% as soon as possible. Blood oxygenation can be increased, in many instances, by altering other ventilatory or perfusion parameters.

11.4. Fluid management

Rapid volume expansion is necessary to correct the hemodynamic abnormalities of severe sepsis. Two large intravenous catheters should be placed in a patient with severe sepsis so that adequate volumes of fluid can be administered quickly if necessary.

Beside whole blood, there are two types of fluid available: crystalloid (no oncotic pressure) or colloid (oncotically active). Red blood cells should be administered if necessary to maintain adequate oxygen-carrying capacity. As previously stated, this usually means a hematocrit around 30–35%.

Both *crystalloid* and *colloid* solutions are effective in restoring circulating volume, the major difference being the increased quantities required with a noncolloid solution such as lactated Ringer's or normal saline (Holcroft et al 1977; Sibbald et al 1983; Virgilio 1979). This also results in some degree of plasma protein depletion. Moderate degrees of depletion are well tolerated by the lungs with an intact endothelium because of the compensatory decrease in interstitial protein content, allowing for the maintenance of a normal oncotic gradient (Demling et al 1979; Harms et al 1982). Also, systemic interstitial protein stores can be mobilized to correct the transient hypoproteinemia and lead to systemic edema formation.

Plasma proteins, of course, do more than generate oncotic pressure. Albumin is the principal binding agent for free fatty acids and a number of other potentially toxic substances. Globulins are necessary for host defense and plasma fibronectin for clearance of microaggregates. The larger proteins may also be responsible for assisting in maintaining the integrity of the interstitial space. Proteins in the form of *fresh-frozen plasma* will restore these safety factors but should be used appropriately because of tremendous expense and limited supply. Albumin solutions can also be used, but are the second choice because the clumping of albumin molecules, which is seen in these solutions, will decrease both the oncotic pressure and binding potential.

Hypertonic crystalloid salt solutions have other added properties. Solutions such as hypertonic saline (Na^+ = 300 meq/liter) have been reported to decrease pulmonary and systemic vascular resistance (Eliakim et al 1958). This may improve pulmonary and systemic blood flow in the septic shock patient. These solutions have mild diuretic properties. Less fluid is required for resuscitation, as part of the intravascular fluid volume deficit is corrected by the removal of cell water because of the increased extracellular osmotic pressure. It remains to be determined whether this degree of depletion of cell water is deleterious to cell function. We have reported that the increase in pulmonary artery pressure which is seen during resuscitation of sheep from hemorrhagic shock is markedly decreased when hypertonic saline is used as resuscitation fluid. Cardiac output is also restored more rapidly (Nerlich et al 1983). Controlled studies in man have only been performed after burn injury where significantly less fluid is required with the use of hypertonic salt solutions. No data are yet available on sepsis.

Dextrans are macromolecules of branched polysaccharides which have a number of properties advantageous for the treatment of hypovolemic shock. First, the dextrans are very large in molecular radius, considerably larger than proteins of comparable molecular weight (Thoren 1978). This is due to their branched rather than globular structure. This larger size increases intravascular retention. Secondly, gram for gram, dextrans have a colloid osmotic pressure over twice that of albumin. Because of their larger size, interstitial dextran concentration remains low, markedly increasing the transvascular colloid osmotic gradient and, consequently, increasing plasma volume (Kramer et al 1982b). Thirdly, these solutions are considerably cheaper than protein solutions.

There are some complications attributed to dextrans. There is a small incidence of anaphylactic reaction to dextran infusion (less than 0.01%). With high molecular solutions, red cell aggregation and increased blood viscosity can occur, and there is difficulty with cross-matching blood for transfusion. Smaller molecular weight solutions such as rheomacrodex (dextran-40) have fewer complications. The lower molecular weight compounds are more rapidly excreted in urine. Although urine viscosity increases dramatically, studies in dogs have revealed no deleterious renal effects. Dextran-40 also improves microcirculatory blood flow by its anti-aggregating properties on red cells and platelets. Interestingly, dextran infusion accentuates the hypoproteinemia seen with shock and resuscitation. This may be due to a decreased return of interstitial protein to the plasma space, as the result of the contracted interstitial space and decrease in lymph flow. It may also partly be due to expanded plasma volume. It is important to recognize that once the dextran infusion has stopped, protein replacement is necessary, or the effects of hypoproteinemia (i.e. edema formation) will occur.

11.5.　Pharmacologic manipulation

11.5.1.　Inotropic agents

When adequate tissue perfusion cannot be obtained with volume alone, the use of vasoactive drugs may be needed (Teule et al 1983; Wilson 1972). In general, vasopressors are contraindicated in that although blood pressure may be improved, microcirculatory blood flow will be further diminished, potentiating ischemia and the release of mediators which will further accentuate the hemodynamic instability. The only rationale for vasopressors is a temporary use to restore aortic blood pressure to a level sufficient to maintain coronary perfusion and prevent further pump failure. Isoproterenol and dopamine are the agents of choice. Both agents have chronotropic and inotropic effects. Dopamine is now the treatment of choice because its major beta adrenergic effect is inotropic (Teule et al 1983). Dopamine in low doses also improves renal perfusion, thereby diminishing the chance of renal dysfunction. Digitalis appears to be a less effective agent when an immediate improvement in cardiac output is needed.

It must again be emphasized that volume restoration is the key to resuscitation, while these other agents should be used only when necessary.

11.5.2.　Corticosteroids

These agents will be covered in Chapter 8.

11.5.3.　Prostaglandin inhibition

Having described an increase in PG production, particularly thromboxane, a number of investigators reported a protective effect of inhibition of prostaglandin synthesis on the endotoxin-induced pulmonary hypertension. Parratt et al (1982) have clearly demonstrated that various prostaglandin inhibitors can completely abolish the initial pulmonary hypertension. We have found a similar effect with endotoxin lung injury in sheep.

A decrease in pulmonary hypertension during the septic injury would improve not only pulmonary function by decreasing capillary hydrostatic pressure, but also cardiac function by decreasing right heart strain and left heart compliance. Several cyclooxygenase inhibitors have been used in animal studies. Ibuprofen is an agent which may have clinical usefulness since the drug is relatively safe in man with minimal side effects. Animal studies have demonstrated its effectiveness in decreasing sepsis-induced hypertension, bronchoconstriction, and hypoxia (Fig. 10). Several studies demonstrated improved survival rates and hemodynamic stability with generalized sepsis in animals (Fletcher and Ramwell 1981). Prostacyclin release may increase vasodilatation of sepsis. Clinical trials are limited by the fact that the intravenous form of the drug has not as yet been released for human studies.

Two potential difficulties arise with the use of prostaglandin inhibitors. This relates to the fact that inhibition of cyclooxygenase activity may shunt arachidonic acid metabolism into the lipoxygenase pathway. The products of this pathway, that is, the leukotrienes, may well accentuate lung injury. This

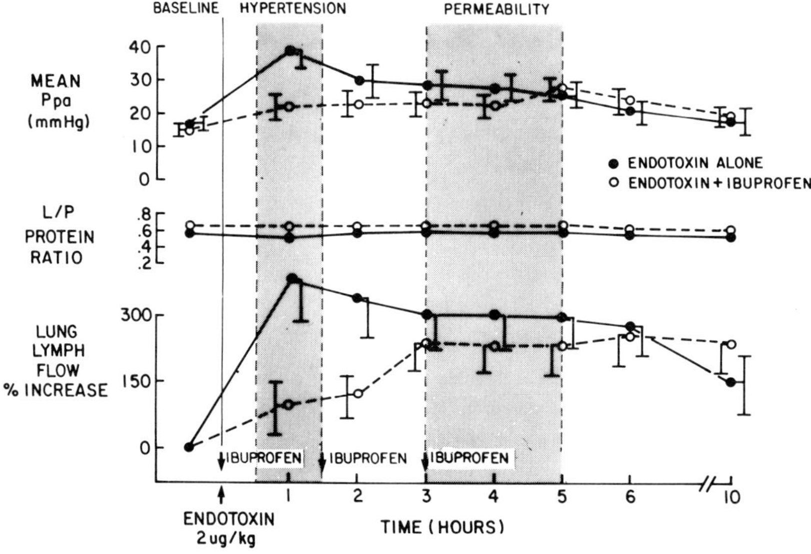

Fig. 10 *The effect of ibuprofen on the endotoxin-induced pulmonary microvascular injury in 3 sheep (mean±SD). The response is compared to a study in the same animals with endotoxin alone. Ibuprofen was given as a bolus in a dose of 12.5 mg/kg just prior to endotoxin. A dose of 12.5 mg/kg was then given at 90 and 180 minutes after endotoxin. As can be seen, the early pulmonary hypertension and lung lymph flow response were decreased with ibuprofen. The early hypoxia and increased respiratory rate response were also decreased. The late permeability phase beginning at 3 hours was not altered. White blood cell count decreased to a comparable degree in both studies.*

process has not been well studied because it is extremely difficult to measure these products. There appear to be major differences in the various inhibitors as to the propensity to shunt metabolism into this pathway. Indometacin clearly does so, whereas ibuprofen does not. The second problem is that cyclooxygenase inhibition not only removes thromboxane but also prostacyclin. The latter agent may protect the lung by maintaining membrane integrity but it accentuates systemic hypotension.

Selective inhibition of thromboxane by inhibition of the enzyme thromboxane synthetase may be more effective. A number of selective inhibitors are in the process of being studied. One such agent, ketoconazole, is already being used in man as an antifungal agent. Systemic hemodynamics have also been reported to be improved in animals with a number of these agents. Less is currently known about the role of prostaglandins on the heart and systemic circulation than has been determined for the lung. There is no good evidence that prostaglandin inhibition will alter the increase in pulmonary capillary permeability seen after endotoxin or sepsis, or the impairment of oxygen utilization seen in the systemic circulation.

Although preliminary data in man indicate that thromboxane may play a role in sepsis-induced injury, conclusive data are needed before the potential benefits of any of these agents can be addressed.

11.5.4. Prostaglandin infusion

A number of animal studies have demonstrated that exogenous prostacyclin (Fig. 11), and, to a lesser extent, PGE_1, decreases the clinical and anatomical manifestations of sepsis-induced increased pulmonary capillary permeability (Demling et al 1981c; Slotman et al 1982). The mechanism of this protective effect remains unclear but may be related purely to its vasodilatory effects, decreasing capillary pressure. Prostacyclin, however, does appear to stabilize the neutrophil, which may in turn attenuate neutrophil-induced injury. There are several clinical trials underway of PGI_2 and PGE_1. However, insufficient data are available to determine their usefulness. PGI_2 is very unstable in solution, and this fact, plus the finding that potency varies considerably between batches of the agent, make long-term clinical usefulness unlikely. There is also good evidence that PGI_2, as well as other vasodilators, impairs hypoxic pulmonary vasoconstriction and can further accentuate a 'shunt' due to sepsis-induced ventilation–perfusion mismatch by inhibiting the normal compensatory mechanisms.

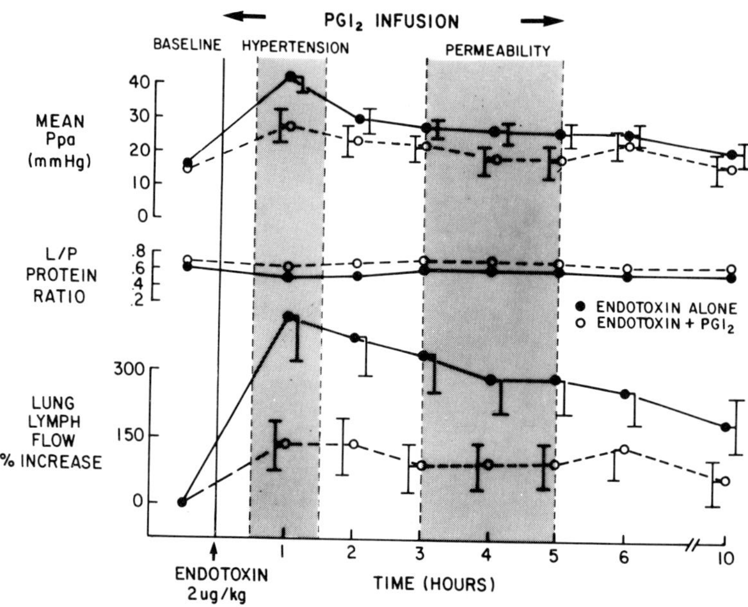

Fig. 11   *The effect of PGI₂ infusion on endotoxin-induced lung injury in 5 sheep (mean±SD). PGI₂ was infused (0.2 µg/kg per minute) for 5 hours beginning at the time of endotoxin injection. The response is compared to a study in the same animals with endotoxin alone. The sequence of the paired studies was randomized. The early pulmonary hypertension and the later moderate hypertension were significantly reduced. The lymph flow response during both phases were also significantly reduced with PGI₂. The L/P protein response was slightly increased in the PGI₂ group compared to endotoxin alone, indicating the presence of an increase in permeability. A transient increase in pulmonary artery pressure and lymph flow was seen after PGI₂ infusion was stopped.*

11.5.5. Lipoxygenase pathway inhibition

There is some preliminary evidence that levels of products of the lipoxygenase pathway are elevated during sepsis in animals. A number of lipoxygenase pathway inhibitors are being studied in sepsis animal models. No clinical date are yet available.

11.5.6. Antioxidants

Superoxide dismutase (SOD) and catalase are cell enzymes which normally detoxify the ever-present oxygen radicals. When these enzymes are decreased in amount or oxygen radicals increased in amount, cell damage results. Infusion of SOD and/or catalase has been found to be effective in decreasing lung injury from hyperoxia in animals. There are scattered case reports of the treatment of paraquat poisoning, which results in an oxygen radical injury in man, with SOD with varied results. There have been a number of recent attempts in animals to infuse these enzymes during sepsis to try and decrease oxygen radicals, assuming that this is, in fact, the cause of the injury. This has not met with much success. The reasons for this may be that the enzymes are rapidly cleared by the kidneys and that the site of action is at the cell and not in the circulating blood. The transfer of the enzyme into the cell through liposomes may be more beneficial.

A number of antioxidants or oxygen radical scavengers are available, particularly against the hydroxyl ($OH^{\cdot}$) radical, which appears to be the most harmful. Although these agents have been shown to be useful in a number of acute injury states due to oxygen radicals such as hyperoxia, no good evidence is yet available that a similar protection will be seen with other forms of ARDS. Many of these antioxidants, such as mannitol, various urea compounds, and vitamin E, are relatively free of serious side effects in man, and therefore clinical usefulness remains a strong possibility. The clearance of the radicals liberated from the neutrophil would prevent tissue damage but should not impair neutrophil killing ability, a distinct advantage over agents such as the corticosteroids.

11.5.7. Antiserotonin agents

Serotonin (5-HT) produces both broncho- and vasoconstriction when released. This agent is rapidly metabolized in the lung. However, if the lung metabolic activity is decreased or serotonin release increased, symptoms will develop. Ketanserin is a serotonin antagonist which inhibits 5-HT activity at the receptor level (Wilson 1972). Clinical trials are underway for the treatment of pulmonary hypertension in man.

11.5.8.   Anticoagulants

The products of coagulation have been reported to contribute to sepsis injury. Heparin therapy has been proposed to decrease the activity of the coagulation cascade and thereby decrease the potentially deleterious byproducts of microembolization (Binder et al 1979; Vaage 1982). Large amounts of heparin are necessary to decrease clotting in these already hypercoagulable patients. The release of tissue thromboplastin and the presence of byproducts of ischemic tissue accelerate clotting. Despite the theoretical advantage of heparin therapy, there are no data of any beneficial effect in sepsis in man and very unconvincing data from animal studies. Bleeding complications of high-dose heparin remain a major concern.

11.5.9.   Fibronectin repletion

The role of both plasma and tissue fibronectin in lung injury from sepsis remains, as yet, undetermined, as previously pointed out. There are several clinical series indicating that correction of a fibronectin deficiency improves lung function in patients with sepsis (Scovill et al 1979). There are also other series which show no correlation between fibronectin levels and lung disease and no evidence of improvement with repletion (Grossman et al 1983). It therefore remains to be determined which patients with fibronectin deficiency will have improved lung function with repletion. The effect of fibronectin repletion on the systemic effects of sepsis are less well studied.

12.   LUNG METABOLISM AND SEPSIS-INDUCED RESPIRATORY FAILURE

Only recently has it been appreciated that the pulmonary endothelial cell has major metabolic as well as respiratory functions (Holcroft et al 1980). It may soon become apparent that the hypoxia manifested when respiratory function is impaired may be only one of many deleterious consequences of severe pulmonary injury. The pulmonary endothelial cell is active in metabolizing a number of biogenic amines, peptides, and, as previously mentioned, prostaglandins. The lung is a major site for degradation of serotonin and, to a lesser degree, a site for degradation of norepinephrine. Metabolism of serotonin occurs in all pulmonary vessels. This function has been found to be impaired after hyperbaric oxygen injury, which results in increasing levels of this systemic vasodilator and pulmonary vasoconstrictor in the blood.

Of greater importance is the role of the lung in peptide metabolism, in particular the degradation of bradykinin and the conversion of angiotensin I to angiotensin II. The major site for both of these activities is in the lung, actually in the same enzyme system. Therefore, extraction of the most potent vasodilator in the body, bradykinin, and activation of the most potent vasoconstrictor, angiotensin II, occur in the lung. Patients with ARDS have been found to have decreased levels of this enzyme, which leads to an increase in circulating bradykinin and a decrease in angiotensin. This may explain the systemic hypotension that can accompany sepsis-induced ARDS. Cardiopulmonary bypass, in which the lung is excluded from the circulation, has been found to result in increased levels of endogenous bradykinin. This is further evidence of the significance of nonrespiratory pulmonary function. The lung is the major site of breakdown of prostaglandins of the E and F series and also a major source of these prostaglandins and prostacyclin. Alterations in this system may well potentiate failure of other organ systems. We may soon see that the description of ARDS also includes these major abnormalities.

13. SUMMARY

Cardiopulmonary failure from sepsis is a complex disease with both circulatory and ventilatory components. The exact pathophysiology has not been determined and, therefore, the appropriate pharmacological manipulation of the injury remains speculative. However, prevention and treatment can be effectively accomplished if the basic alterations of the normal physiologic state are recognized and basic principles applied to correct these abnormalities.

REFERENCES

Aasen AO, Frolich W, Saugstad OD, Amundsen E (1978) Plasma kallikrein activity and pre-kallikrein levels during endotoxin shock in dogs. *Eur. Surg. Res. 10*, 50-62.
Adams T, Traber D (1972) The effects of the prostaglandin synthetase inhibitor

Ibuprofen on the cardiopulmonary response to endotoxin in sheep. *Circ. Shock* 9, 481-489.

Anderson FL, Juberg W, Fralios A, Tsagoris T, Kiuda H (1972) Plasma prostaglandin levels during endotoxin shock in dogs. *Circulation 45*, 124-130.

Anderson RR, Holliday RL, Driedger AA, Sibbald W (1979) Documentation of pulmonary capillary permeability in the adult respiratory distress syndrome accompanying human sepsis. *Am. Rev. Resp. Dis. 119*, 869-877.

Baudendistel L, Shields J, Kaminski D (1982) Comparison of double indicator thermodilution measurements of extravascular lung water (EVLW) with radiographic estimation of lung water in trauma patients. *J. Trauma 22*, 938-988.

Bedrossean CWM, Woo J, Miller WC (1979) Decreased angiotensin-converting enzyme in the adult respiratory distress syndrome. *Am. J. Clin. Pathol. 70*, 244-247.

Binder AS, Nakahara K, Ohkuda K, Kageler W, Staub NC (1979) Effect of heparin or fibrinogen depletion on lung fluid balance in sheep after emboli. *J. Appl. Physiol. 47*, 213-219.

Blain CM, Anderson TO, Pietras R, Gunnar R (1970) Immediate hemodynamic effects of gram negative vs. gram positive bacteremia in man. *Arch. Intern. Med. 126*, 260-265.

Blaisdell FW (1981) The role of steroids in septic shock. *Circ. Shock 8*, 673-682.

Bodai B, Blaisdell FW (1981) Discontinuance of positive end expiratory pressure during suctioning: effect of oxygenation in patients with adult respiratory distress syndrome. *Surg. Forum 31*, 311-312.

Boogaerts M, Yamada O, Jacob H, Moldow C (1982) Enhancement of granulocyte-endothelial adherence and granulocyte-induced cytotoxicity by platelet release products. *Proc. Natl Acad. Sci. USA 79*, 7019-7023.

Brigham KL, Bowers R, Haynes J (1979) Increased sheep lung vascular permeability caused by *Escherichia coli* endotoxin. *Circ. Res. 45*, 292-297.

Brigham KL, Kariman K, Harris T (1983) Correlation of oxygenation with vascular permeability-surface area but not with lung water in humans with acute respiratory failure and pulmonary edema. *J. Clin. Invest. 72*, 339-349.

Busch C, Dahlgren S, Jacobsson S, Jung B, Modig J, Saldeen T (1971) Determination of fibrin trapping in the lungs of patients developing the microembolism syndrome. *Acta Chir. Scand. 137*, 599-601.

Busch C, Rommer L, Saldeen T (1973) Quantitation of fibrin deposition and elimination in organs of rats injected with labelled fibrinogen. *Thromb. Diath. Haemorrh. 29*, 94-100.

Busch C, Lindquist O, Saldeen T (1974) Respiratory insufficiency in the dog induced by pulmonary microembolism and inhibition of fibrinolysis. *Acta Chir. Scand. 140*, 255-266.

Calvin JE, Driedger A, Sibbald WJ (1981) Positive end-expiratory pressure (PEEP) does not depress left ventricular function in patients with pulmonary edema. *Am.*

Rev. Resp. Dis. 124, 121-128.

Cann M, Stevenson T, Fiallos E (1972) Depressed cardiac performance in sepsis. *Surg. Gynecol. Obstet. 134*, 759-766.

Clements JH (1967) The alveolar lining layer. In: DeReuch AVS (Ed), *Development of the lung*, pp 202-237. Ciba Foundation Symposium.

Clowes G, Farrington G, Zieschnerd W, Suschneid W (1970) Circulatory factors in the etiology of pulmonary insufficiency and right heart failure accompanying severe sepsis (peritonitis). *Ann. Surg. 171*, 663.

Clowes GH, Hirsch E, Williams L, Weidner M (1974) Septic lung and shock lung in man. *Ann. Surg. 181*, 681-691.

Coalson JJ, Hinshaw LB, Guenther CA (1970) The pulmonary ultrastructure in septic shock. *Exp. Mol. Pathol. 12*, 84-103.

Cochrane CG, Spragg R, Revak SD (1970) Pathogenesis of the adult respiratory distress syndrome: evidence of oxidant activity in bronchoalveolar lavage fluid. *J. Clin. Invest. 71*, 754-761.

Comper WD, Laurent TC (1978) Physiological function of connective tissue polysaccharides. *Physiol. Rev. 58*, 255-315.

Comroe JH, Forster RE (1968) The lung: clinical physiology and pulmonary function tests. In: *Chicago Yearbook Medical Publications, 2nd Ed.*

Craddock PR, Fehr J, Brigham KL, Kronenberg RS, Jacob HS (1977) Complement and leukocyte mediated pulmonary dysfunction in hemodialysis. *N. Engl. J. Med. 295*, 796-774.

Crapo JD, Tierney DF (1974) Superoxide dismutase and pulmonary oxygen toxicity. *Am. J. Physiol. 226*, 1401-1407.

Demling RH (1981) The pathogenesis of respiratory failure after trauma and sepsis. *Surg. Clin. North Am. 60*, 1373-1390.

Demling RH (1982) Role of prostaglandins in acute microvascular injury. *Ann. N.Y. Acad. Sci. 384*, 517-534.

Demling RH, Manohar M, Belzer FO (1979) Effect of plasma oncotic pressure on the pulmonary microcirculation after hemorrhagic shock. *Surgery 86*, 323-332.

Demling RH, Proctor R, Duy N, Starling JR (1980) Lung lysosomal enzyme release during hemorrhagic shock and endotoxemia. *J. Surg. Res. 28*, 269-279.

Demling RH, Smith M, Gunther R, Flynn J, Gee M (1981a) Pulmonary injury and prostaglandin production during endotoxemia in conscious sheep. *Am. J. Physiol. 240*, 348-353.

Demling RH, Smith M, Gunther R, Wandzilak T (1981b) Endotoxin-induced lung injury in unanesthetized sheep: effect of methylprednisolone. *Circ. Shock. 8*, 351-360.

Demling RH, Smith M, Gunther R, Gee M, Flynn J (1981c) The effect of prostacyclin infusion on endotoxin-induced lung injury. *Surgery 89*, 257-263.

Duff H, McLean P, MacLean D (1969) Defective oxygen consumption in septic shock. *Surg. Gynecol. Obstet. 129*, 1051-1060.

Eliakim M, Rosenberg SI, Braun K (1958) Effect of hypertonic saline of the pulmonary and systemic pressures. *Circ. Res. 6*, 357-362.

Fein H, Grossman RF, Jones JG (1979) The value of edema fluid protein measurement in patients with pulmonary edema. *Am. J. Med. 62*, 32-38.

Finley RJ, Holliday RL, Lefcoe M, Duff JH (1975) Pulmonary edema in patients with sepsis. *Surg. Gynecol. Obstet. 140*, 851-857.

Fletcher JR, Ramwell PW (1981) Indomethacin treatment following baboon endotoxin shock improves survival. In: Shumer WF (Ed), *Advances in Shock Research*, pp 103-111. Alan Liss, New York.

Flick MR, Perel A, Kageler W, Staub NC (1979) White blood cells contribute to increased lung vascular permeability after microemboli. *Fed. Proc. 38*, 1327.

Flynn JT, Demling RH (1982) Inhibition of endogenous thromboxane synthesis by exogenous prostacyclin during endotoxemia in conscious sheep. *Adv. Shock. Res. 7*, 199-207.

Giordino J, Zinner M, Hobson RW, Gerven A (1976) The effect of microaggregates in stored blood on canine pulmonary vascular resistance. *Surgery 80*, 617-623.

Goldfarb RD, Wiber P, Eisenman J (1979) Isolation of a shock-induced circulatory cardiodepressant substance. *Am. J. Physiol. 237*, 168-174.

Greenfield LJ, Jackson R, Elkins R (1974) Cardiopulmonary effects of volume loading of primates in endotoxin shock. *Surgery 76*, 560-569.

Greenwald RA, Moy WW (1980) Effect of oxygen-derived free radicals on hyaluronic acid. *Arthritis Rheum. 23*, 455-463.

Grossman J, Will L, Hahn N, Exten R (1983) A randomized trial of cryoprecipitate therapy in critically ill patients: preliminary results. *Am. Rev. Resp. Dis. 127*, 127.

Hammerschmidt DE, Hudson LD, Weaver LJ, Craddock PR, Jacob HS (1980) Association of complement activation and elevated plasma $C5_a$ with adult respiratory distress syndrome. *Lancet 1*, 947-949.

Hardaway RM (1973) Disseminated intravascular coagulation as a possible cause of acute respiratory failure. *Surg. Gynecol. Obstet. 137*, 419-423.

Harms B, Bodai B, Kramer G, Demling RH (1982) Microvascular transport of fluid and protein in pulmonary and systemic circulation after thermal injury. *Microvasc. Res. 23*, 77-86.

Hechtman HB, Lonergan EA, Staunton PB, Dennis R, Shepro D (1978) Pulmonary entrapment of platelets during acute respiratory failure. *Surgery 83*, 277-283.

Heflin AC, Brigham KL (1981) Prevention by granulocyte depletion of increased vascular permeability of sheep following endotoxin. *J. Clin. Invest. 68*, 1253-1260.

Heidman M (1979) Complement activation in vitro induced by endotoxin and injured tissue. *Surg. Res. 26*, 670-674.

Hess ML, Hastillo A, Greenfield LJ (1981) The cardiovascular spectrum of gram-negative shock. *Prog. Cardiovasc. Dis. 13*, 279-299.

Hinshaw LB, Archer LT, Black MR (1974) Myocardial function in endotoxin shock. *Am. J. Physiol. 226*, 357-363.

Hinshaw LB, Solomon LA, Freeny PC (1967) Endotoxin shock: hemodynamic and survival effects of methylprednisolone. *Arch. Surg. 94*, 61-66.

Holcroft J, Trunkey D, Carpenter M (1977) Sepsis in the baboon: factors affecting resuscitation and pulmonary edema in animals resuscitated with ringer's lactate versus plasmanate. *J. Trauma 17*, 600-610.

Holcroft J, Trunkey D, Carpenter M (1979) Extravasation of albumin in tissues of normal and septic baboons and sheep. *J. Surg. Res. 26*, 341-347.

Holcroft J, Trunkey D, Carpenter M (1980) Extracellular calcium pool decreases during deep septic shock in the baboon. *Ann. Surg. 5*, 683-686.

Kaplan RL, Sahn SA, Petty TL (1979) Incidence and outcome of the respiratory distress syndrome in gram negative sepsis. *Arch. Intern Med. 139*, 867-869.

Kinney JM, Askanazi J, Gump FE (1980) Use of ventilatory equivalent to separate hypermetabolism from increased dead space ventilation in the injured or septic patient. *J. Trauma 20*, 111-191.

Klebanoff SJ (1980) Oxygen metabolism and the toxic properties of phagocytes. *Ann. Intern. Med. 93*, 480-489.

Kramer GC, Harms BA, Gunther R, Renkin E, Demling RH (1981) The effects of hypoproteinemia on blood-to-lymph fluid transport in sheep lung. *Circ. Res. 49*, 1173-1181.

Kramer GC, Harms BA, Demling RH, Renkin EM (1982a) Mechanisms of redistribution of plasma protein following acute protein depletion. *Am. J. Physiol. 243*, 803-869.

Kramer GC, Gunther R, Nerlich M, Zwerfach S, Demling RH (1982b) Effect of dextran-70 on increased microvascular fluid and protein flux after thermal injury. *Circ. Shock 9*, 529-541.

Krausz M, Perel A, Eimerl D, Coter S (1977) Cardiopulmonary effects of volume loading in patients in septic shock. *Ann. Surg. 183*, 429-434.

Kuchelt W, Dauberschmidt R, Scharfenbert J (1980) Application of bromhexine metabolite VII and bromhexine-glucose in the treatment of adult respiratory distress syndrome. *Respiration 29*, 264-271.

Kux M, Coalson JJ, Masscon WH (1972) Pulmonary effects of *E. coli* endotoxin: role of leukocytes and platelets. *Ann. Surg. 175*, 26-34.

Lava J, Rice C, Moss G, Levine H (1982) Pulmonary dysfunction in sepsis: is pulmonary edema the culprit? *J. Trauma 22*, 280-284.

Lee CT, Fein AM, Lippman M, Holtzman H (1981) Elastolytic activity in pulmonary lavage fluid from patients with adult respiratory distress syndrome. *J. Clin. Invest. 304*, 192-196.

Lewis FR, Blaisdell FW, Scholbohm RM (1977) Incidence and outcome of post-traumatic respiratory failure. *Arch. Surg. 112*, 436-443.

Lewis FR, Elings VB, Sturm JA (1979) Bedside measurement of lung water. *J. Surg. Res. 27*, 250-261.

Malik AB, Johnson A, Tahamont M (1982) Mechanism of lung vascular injury after intravascular coagulation. *Ann. N.Y. Acad. Sci. 384*, 213-235.

Martyn JX, Snider MT, Farago LF, Burke JF (1981) Thermodilution right ventricular volume: a novel and better predictor of volume replacement in acute thermal injury. *J. Trauma 21*, 619-626.

Maunder RJ, Harlan JM, Carrico CJ (1983) Fibronectin levels in patients developing ARDS. *Am. Rev. Resp. Dis. 127*, 95.

McCraig DJ, Parratt JR (1980) Reduced myocardial response to calcium during endotoxin shock in the cat. *Circ. Shock 7*, 23-30.

McCraig D, Kane K, Bailey G (1979) Myocardial performance in feline endotoxin shock: a correlation between myocardial contractility, electrophysiology and ultra structure. *Circ. Shock 6*, 201-210.

McDonald JA, Baum BJ, Rosenberg DM, Kelman J, Brin S, Crystal R (1979) Destruction of a major extracellular adhesive glycoprotein (fibronectin) of human fibroblasts by neural proteases from polymorphonuclear leukocyte granules. *Lab. Invest. 40*, 350-357.

Meyer MM (1973) The complement system. *Science 229*, 54-66.

Milligan GF, MacDonald J, Mellon A, Ledingham I (1974) Pulmonary and hematologic disturbances during septic shock. *Surg. Gynecol. Obstet. 138*, 43-49.

Moncada S, Vane JR (1979) Arachidonic acid metabolites and the interactions between platelets and blood vessel walls. *N. Engl. J. Med. 300*, 1142-1147.

Moss G, Staunton C, Stein AA (1973) The centrineurogenic etiology of the adult respiratory distress syndrome. *Am. J. Surg. 126*, 37-51.

Nerlich M, Flynn J, Demling RH (1983) Effect of thermal injury on endotoxin-induced lung injury. *Surgery 93*, 289-296.

Nerlich M, Gunther R, Demling RH (1983) Resuscitation from hemorrhagic shock with hypertonic saline or lactated ringer's; effect on the pulmonary and systemic microcirculation. *Circ. Shock 10*, 179-183.

O'Flaherty JT, Showell HJ, Ward PA, Showell H (1979) A possible role of arachidonic acid in human neutrophil aggregation and degranulation. *Am. J. Pathol. 96*, 799-809.

Ogletree ML, Brigham KL, Oates JA, Hubbard WC (1981) Increased flux of 5-HETE in sheep lung lymph during pulmonary leukostasis after endotoxin. *Fed. Proc. 40*, 3053.

Parratt J, Coker S, Hughes B, MacDonald A (1982) The possible role of prostaglandins and thromboxane in the pulmonary consequences of experimental endotoxic shock and clinical sepsis. In: McConn R (Ed), *Role of Chemical Mediators in the Pathophysiology of Acute Illness and Injury*, pp 195-219. Raven Press, New York.

Parwitt RM, Ghignone M (1983) Treatment of right ventricular dysfunction in acute respiratory failure. *Crit. Care Med. 11*, 346-353.

Petrini M, Peterson BT, Hyde RW (1978) Lung tissue volume and blood flow by rebreathing: theory. *J. Appl. Physiol. 44*, 795-802.

Petty TL, Silvers GW, Paul GW (1979) Abnormalities in lung elastic properties and surfactant function in adult respiratory distress syndrome. *Chest 75*, 571-574.

Pietra GG, Ruttner JR, Wust W, Glinz W (1981) The lung after trauma and shock – fine structure of the alveolar capillary barrier in 23 autopsies. *J. Trauma 21*, 454-462.

Raffa J, Trunkey DD (1978) Myocardial depression in sepsis. *J. Trauma 18*, 617-624.

Reynolds DC, Gurll NJ, Vargish T (1980) Blockade of opiate receptors with naloxone improves survival and cardiac performance in canine endotoxin shock. *Circ. Shock 7*, 39-46.

Robertson B (1978) Surfactant substitution: experimental models and clinical applications. *Lung 158*, 57-68.

Saba TM, Blumenstock FA, Weber P, Kaplan J (1979) Physiologic role of cold-insoluble globulin in systemic host defense: implications of its characterization as the opsonic α_2 surface binding glycoprotein. *Ann. N.Y. Acad. Sci. 312*, 43-55.

Sachs T, Moldow CF, Craddock PR, Bowers T, Jacobs H (1978) Oxygen radicals mediate endothelial damage by complement-stimulated granulocytes. *J. Clin. Invest. 60*, 1161-1167.

Saldeen T (1982) Fibrin derived peptides and pulmonary injury. *Ann. N.Y. Acad. Sci. 384*, 319-332.

Samuelsson B, Borgeat B, Hammarstrom S, Murphy R (1980) Leukotrienes: a new group of biologically active compounds. *Adv. Prostaglandin Thromboxane Res. 6*, 1-18.

Scovill WA, Saba TM, Blumenstock FA, Bernard H, Powers S (1979) Opsonic α_2 surface binding glycoprotein therapy during sepsis. *Ann. Surg. 188*, 521-529.

Sibbald W, Driedger A (1983) Right ventricular function in acute disease states: pathophysiologic considerations. *Crit. Care Med. 11*, 358-364.

Sibbald WJ, Driedger AA, Wells GA, Myers M, Lefcoe M (1983) The short term effects of increasing plasma colloid pressure in patients with noncardiac pulmonary edema. *Surgery 93*, 620-633.

Siegel JH, Greenspan M, El Guercio L (1967) Abnormal vascular tone, defective oxygen transport and myocardial failure in human septic shock. *Ann. Surg. 165*, 504-513.

Sjorstrand U (1980) High frequency positive pressure ventilation, HFPPV: a review. *Crit. Care Med. 8*, 345.

Slotman G, Machiedo G, Casey K, Lyons M (1982) Histologic and hemodynamic effects of prostacyclin and prostaglandin E_1 following oleic acid infusion. *Surgery 92*, 93-99.

Smith ME, Gunther R, Gee M, Flynn J, Demling RH (1981a) Leukocytes, platelets, and thromboxane A_2 in endotoxin-induced lung injury. *Surgery 90*, 102-107.

Smith M, Gunther R, Zaiss C, Flynn J, Demling RH (1981b) Pulmonary microvascular injury from lipoxygenase infusion: comparison with endotoxemia. *Circ. Shock 8*, 647-656.

Smith M, Gunther R. Zaiss C, Demling RH (1982) Prostaglandin infusion and endotoxin-induced lung injury. *Arch. Surg. 117*, 175-181.

Snow RL, Davies P, Pontoppidan H, Zapol W, Reid L (1982) Pulmonary vascular remodeling in adult respiratory syndrome. *Am. Rev. Resp. Dis. 126*, 887-892.

Staub NC (1974) Pulmonary edema. *Physiol. Rev. 54*, 678-711.

Sugarman H, Strash A, Hirsch J, Gauser F, Shirazi K, Sharp D, Greenfield L (1981) Sensitivity of scintigraphy for detection of pulmonary capillary albumin lead in canine oleic acid ARDS. *J. Trauma 21*, 520-527.

Tate RM, Van Benthuysen KM, Shasby DM (1981) Dimethylthiomea, a hydroxyl radical scavenger, blocks oxygen radical-induced acute edematous lung injury in isolated perfused lung. *Am. Rev. Resp. Dis. 123*, 243.

Tatum J, Burke S, Sugarman H, Strash S, Greenfield L (1982) Computerized scintigraph techniques for the evaluation of adult respiratory distress syndrome: initial clinical trials. *Nucl. Med. 143*, 238-242.

Thoren L (1978) Dextran as a plasma volume substitute. In: *Blood Substitutes and Plasma Expanders*, pp 265-282. Allan Liss, New York.

Teule GJ, Den Hollander W, Bronsveld W (1983) Effect of volume loading and dopamine on hemodynamics and red cell distribution in canine endotoxin shock. *Circ. Shock 10*, 41-51.

Todd TR, Baile JE, Hogg WC (1978) Pulmonary capillary permeability during hemorrhagic shock. *J. Appl. Physiol. 24*, 298-306.

Tranbaugh R, Elings V, Christensen J, Lewis FR (1982) Determinant of pulmonary interstitial fluid accumulation after trauma. *J. Trauma 22*, 820-826.

Tremper KK, Shoemaker WC (1981) Transcutaneous oxygen monitoring of critically ill adults, with and without flow shock. *Crit. Care Med. 9*, 706-710.

Vaage J (1982) Intravascular platelet aggregation and pulmonary injury. *Ann. N.Y. Acad. Sci. U.S.A. 384*, 301-319.

Vaccaro CH, Rounds S, Hill NS, Brody JS (1982) Molecular charge, pulmonary vascular permeability and capillary basement membrane. *Am. Rev. Resp. Dis. 125*, 279a.

Virgilio RW (1979) Crystalloid vs colloid osmotic resuscitation: is one better? *Surgery 85*, 129-139.

Vincent JL, Degaute JP, Domb M, Simon P, Berre J, Vandersteene A (1983) Serotonin antagonist administration in ARDS. *Am. Rev. Resp. Dis. 127*, 99.

Vito L, Dennis R, Weisel R, Hechtman H (1974) Sepsis presenting as acute respiratory insufficiency. *Surg. Gynecol. Obstet. 138*, 896-901.

Wedmare CV, Williams TJ (1981) Control of vascular permeability by polymorphonuclear leukocytes in inflammation. *Nature 289*, 646-650.

Weinstein ME, Skillman JJ (1981) Management of severe respiratory failure. *Surg. Clin. North Am. 60*, 1403-1412.

Weissman G, Smolen JE, Korchak HM (1980) Release of inflammatory mediators from stimulated neutrophils. *N. Engl. J. Med. 303*, 27-34.

Wilson JW (1972) Treatment or prevention of pulmonary cellular damage with pharmacological doses of corticosteroids. *Surg. Gynecol. Obstet. 134*, 675.

Winslow F, Loeb H, Rahimtoola S (1973) Hemodynamic studies and results of therapy in 50 patients with bacteremic shock. *Am. J. Med. 54*, 421-432.

Wong C, Flynn J, Demling RH (1984) Role of oxygen radicals in endotoxin-induced lung injury. *Arch. Surg. 119*, 77-82.

Zapol WM, Kobayashi K, Snider MT, Green R, Saver M (1977) Vascular obstruction causes pulmonary hypertension in severe acute respiratory failure. *Chest 71 (Suppl)*, 306-307.

Zapol WM, Trelstad RL, Coffey JW, Green R (1979) Pulmonary fibrosis in severe acute respiratory failure. *Am. Rev. Resp. Dis. 119*, 547-554.

Zimmerman G, Renzetti A, Hill N (1983) Functional and metabolic activity of granulocytes from patients with ARDS. *Am. Rev. Resp. Dis. 127*, 290-300.

Handbook of Endotoxin, Vol. 4: Clinical Aspects of Endotoxin Shock
R.A. Proctor, editor
© Elsevier Science Publishers B.V., 1986

CHAPTER 7

Role of antibiotics in endotoxin shock

SIGURDUR GUDMUNDSSON AND WILLIAM A. CRAIG

1. INTRODUCTION

A discussion of the role of antibiotics in endotoxin shock would normally focus on the effects of antimicrobial therapy on gram-negative bacteremia and shock. This topic has been the subject of several reviews published over the past few decades (Barnett and Sanford 1969; Gleckman and Esposito 1981; Hassen 1973; Landesman and Gorbach, 1978; McCabe 1973, 1975; McCabe and Olans 1981; Young et al 1977; Young 1982). However, relatively few clinical studies on the therapy of gram-negative bacteremia have separately evaluated patients in shock. Even less emphasis has been placed on the role of antimicrobials in the prevention of shock. Thus, our discussion and critical review of the available literature cannot be limited entirely to patients in shock. Other questions besides simply the efficacy of antimicrobial therapy in gram-negative bacteremia need to be addressed. Do some antibiotics have a direct inhibitory effect on endotoxins? Do some antibiotics enhance the release of endotoxins and alter the clinical outcome? And finally, do endotoxins influence the absorption, distribution or elimination of antimicrobials? The current review will examine the more limited information on these questions as well.

2. DIRECT INHIBITORY EFFECT OF ANTIMICROBIALS ON ENDOTOXIN ACTION

A variety of in vitro and in vivo model systems have been used to evaluate the ability of antimicrobials to neutralize endotoxin. These include the Limulus lysate assay (Cooperstock 1974), hemagglutination- and hemolysis-inhibition assays (Neter et al. 1958), mouse and chick-embryo lethality (Rifkind and Palmer 1966; Rifkind 1967a, 1967b) and disseminated intravascular coa-

238

gulation and leukopenia in rabbits (Corrigan and Bell 1971a, 1971b). Only the polymyxin class of antibiotics, including polymyxin B, colistin and tyrocidine, have consistently demonstrated an anti-endotoxic effect. Neter et al (1958) were the first to demonstrate the inhibitory effect of polymyxin B on bacterial endotoxin. They also reported that neomycin, paromomycin and streptomycin were effective in preventing endotoxin-induced hemagglutination and hemolysis of sheep erythrocytes. However, the aminoglycoside antibiotics were 12–200 times less active on a weight basis than polymyxin B. Other investigators using different experimental systems have failed to observe any inhibitory effects of the aminoglycoside antibiotics on endotoxin action. Other antimicrobials that have been inactive in these model systems include penicillin G, ampicillin, carbenicillin, sulfadiazine, sulfamethoxazole, erythromycin, tetracycline, chloramphenicol, novobiocin, capreomycin, viomycin, bacitracin, gramicidin, and amphotericin B. Therefore, further discussion will be limited to the polymyxin antibiotics.

2.1 In vitro effects

Polymyxins are cationic branched cyclic decapeptides produced by *Bacillus polymyxa*. They exert their antimicrobial action on gram-negative bacilli by binding to membrane phospholipids and thus disrupting the integrity of the cell membrane (Horton and Pankey 1982; Newton 1956; Sud and Feingold 1972). Electronmicroscopic studies have suggested that polymyxin also disrupts the typical branched-chain structure of endotoxin polymers into smaller fragments or completely disaggregated material (Lopes and Inniss 1969). Polymyxins have further been shown to form complexes with endotoxin by both ionic and hydrophobic interactions (Bader and Teuber 1973; Morrison and Jacobs 1976a) The lipid A 2-keto-3-deoxyoctonate (KDO; 3-deoxy-D-*manno*-2-octulosonic acid) region of the lipopolysaccharide molecule appears to be the binding site for the polymyxins. Lipid A has negatively charged pyrophosphate groups and contains several hydrophobic long-chain fatty acids as well (Rietschel et al 1972; see also Volume 1). The absence of demonstrable anti-endotoxin activity by other cationic compounds such as capreomycin, viomycin and protamine, suggests that a specific stereochemical configuration may be required for endotoxin neutralization (Rifkind and Palmer 1966).

The various in vitro model systems in which the polymyxins have neutralized or reduced endotoxin activity are summarized in Table 1. The largest variety of endotoxins have been studied with the Limulus lysate assay.

TABLE 1   *Model systems in which polymyxin antibiotics have neutralized or reduced endotoxin activity*

Model system	Source of endotoxin	Reference
In vitro:		
Limulus lysate	*E. coli, K. pneumoniae, P. mirabilis, P. aeruginosa, B. pertussis, N. meningitidis, H. influenzae, Y. pestis*	Butler and Möller 1977; Cooperstock 1974; Cooperstock and Riegle 1981
Complement activation	*P. aeruginosa, E. coli, S. sonnei, S. typhimurium, S. minnesota*	Inada 1980; Morrison and Jacobs 1976b; Neter et al 1958
Hemagglutination	*E. coli, S. minnesota*	Neter et al 1958; Warren 1982
Mitogenesis response	*E. coli, S. typhimurium, S. enteritidis, Y. pestis*	Axelrod and Shands 1977, Butler and Möller 1977; Jacobs and Morrison 1975, 1977; Warren 1982
Neutrophil chemotaxis	*E. coli*	Issekutz and Biggar 1978
Biologic effects on neutrophils	*E. coli, S. typhi, S. paratyphi A, B, S. schottmülleri*	Bannatyne et al 1977; Corrigan et al 1974
In vivo:		
Chick embryo	*E. coli, A. aerogenes*	Gruninger and Howe 1969; Rifkind and Palmer 1966; Rifkind 1967b
Mouse survival	*E. coli*	Craig et al 1974; Rifkind 1967a
Generalized and local Shwartzman reaction in rabbits	*E. coli*	Bergstein and Michael 1973; Corrigan and Bell 1971a, 1971b; Craig et al 1974; Rifkind and Hill 1967
Shock in dogs and cats	*E. coli*	From et al 1979; Hughes et al 1981; Palmer and Rifkind 1974

TABLE 1   *(cont'd)*

Model system	Source of endotoxin	Reference
Biliary flow in pigs	*E. coli*	Escartín et al 1982
Fever in goats and rabbits	*E. coli*	Butler and Möller 1977; Van Miert and Van Duin 1977, 1978
Neutrophil count in humans	*S. typhi, S. paratyphi A, B, S. schottmülleri*	Corrigan et al 1974
Animal infections:		
Mice	*S. marcescens, N. meningitidis*	Bannatyne and Cheung 1977, 1979
Rabbits	*P. multocida*	Corrigan and Kiernat, 1979

Polymyxin B and colistimethate deactivated standard endotoxin (*Escherichia coli* O26:B6, Brown preparation) at least 100-fold. Endotoxins from seven other gram-negative organisms were also inhibited but showed marked variation in susceptibility to polymyxin B. In contrast, endotoxin from *Bacteroides fragilis* and *Bacteroides vulgatus* were highly resistant to polymyxin B neutralization. The endotoxin most susceptible to this antibiotic was obtained from a *Proteus mirabilis* isolate, resistant to the antibacterial action of polymyxin B. In addition, calcium did not affect endotoxin neutralization in this study, but is known to reduce antimicrobial activity (Davis et al 1971; Sud and Feingold 1975). This suggests that the bactericidal and endotoxin-inactivating properties of polymyxin B are not directly related.

Polymyxin B inhibits the endotoxin-mediated stimulation of both the classic and alternative complement pathways. Low doses of this antibiotic also inhibit the mitogenic response of lipopolysaccharide in B cells but have no apparent effect on the ability of hapten-conjugated endotoxin to initiate a T cell-independent immune response in vitro (Jacobs and Morrison 1975). Polymyxin B inhibits many of the endotoxin effects on neutrophils including enhancement of nitroblue tetrazolium dye reduction, release of lysosomal enzymes and stimulation of the hexose-monophosphate shunt pathway. Furthermore, the drug neutralizes, at least partly, the endotoxin-mediated suppression of neutrophil chemotaxis.

2.2 In vivo effects

A number of in vivo studies have been performed to examine the inhibitory effect of polymyxin B on the various biologic actions of endotoxin (Table 1). Chick embryo and mouse lethality, the local and generalized Shwartzman reaction in rabbits, shock in dogs and cats, and fever in rabbits and goats have been significantly reduced by the polymyxin group of antibiotics. Even the endotoxin-induced reduction of biliary flow in pigs was reversed by colistin. However, the timing of the polymyxin administration in these studies proved to be critical. In general, the drug needed to be administered relatively close in time to endotoxin challenge. Delay in administration of polymyxin B by 15 to 60 minutes resulted in loss of protection (Corrigan and Bell 1971a; From et al 1979; Gruninger and Howe 1969). This occurs because endotoxin is rapidly cleared from the circulation into tissues (Braude et al 1955).

Prior administration of polymyxin B has also been studied. In chick embryos, drug administered within 6 hours before endotoxin challenge prevented lethality; challenge with endotoxin at 24 hours after polymyxin B administration was not effective (Gruninger and Howe 1969). Antibiotic infusion in dogs as short as 30 minutes prior to endotoxin challenge did not diminish mortality (From et al 1979). On the other hand, Craig et al (1974) studied the effect of pretreatment with multiple doses of polymyxin B which resulted in drug accumulation in tissues, but not in serum. Localization of drug in tissues reduced mouse lethality and the frequency of the generalized Shwartzman reaction in rabbits. However, large doses of polymyxin B (20 mg/kg/d) administered for 2–4 days were required. Polymyxin B pretreatment was ineffective in preventing canine shock and mortality because of the inability of the dog to tolerate high doses of the drug without the relatively rapid development of toxicity.

The quantity of endotoxin neutralized in these animal models by 1 µg of polymyxin B has varied from 2 to 9 ng (Gruninger and Howe 1969; Rifkind and Palmer 1966; Rifkind 1967a). This is in marked contrast to several of the in vitro model systems in which 1 µg of polymyxin was capable of neutralizing from more than 100 to 18 000 ng of endotoxin (Inada 1980; Jacobs and Morrison 1977; Morrison and Jacobs 1976b; Neter et al 1958). In vitro binding studies have suggested that one monomer of polymyxin B combines with one monomer of lipopolysaccharide (Morrison and Jacobs 1976a). The marked deviation from this 1:1 interaction in the in vivo models may be explained by the high tissue binding of polymyxin or the polymyxin–endotoxin complex.

2.3 Animal infection models

A few studies have examined the effects of polymyxin on natural endotoxemia in animals infected with organisms resistant to the drug. A marked decrease in mortality was noted in mice infected intraperitoneally with *Neisseria meningitidis* or *Serratia marcescens* whether polymyxin B was administered 1.5 hours before or 3 hours after onset of infection (Bannatyne and Cheung 1977, 1979). Similarly, treatment with polymyxin B 90 minutes before intraperitoneal injection of *Pasteurella multocida* into rabbits markedly reduced plasma endotoxin and prevented leukopenia and thrombocytopenia (Corrigan and Kiernat 1979). Blood cultures were consistently positive in both groups of rabbits. Polymyxin B given 6 hours after intraperitoneal challenge with *P. multocida* reduced mortality at 24 hours (86% *vs* 100%) and endotoxin levels, but had no significant effect on leukopenia and thrombocytopenia. In additional studies, rabbits received penicillin or penicillin plus polymyxin. The addition of polymyxin B reduced mortality following penicillin therapy from 70 to 50%, but this difference was not significant.

Natural endotoxemia could also result from absorption of endotoxin from the gut. Mucosal damage, mesenteric hypotension or ischemia and the absence of bile flow increase absorption of endotoxin (Cuevas and Fine 1971; Gans and Matsumoto 1974; Kocsar et al 1969). In rats made jaundiced by common bile duct ligation and then challenged with intragastric endotoxin, polymyxin B infusion significantly improved survival from 29% to 73% (Ingoldby 1980). Polymyxin also reduced the degree of endotoxemia in rats following ligation of the mesenteric artery but the drug did not improve survival. However, instillation of non-absorbable antibiotics (gentamicin and carbenicillin) into the intestine of these rats markedly improved survival. Thus, elimination of gram-negative bacteria from the fecal flora and reduction of the endotoxin pool in the gut represents another means in which antibiotics can affect endotoxemia. Studies in rabbits using superior mesenteric artery ligation, bradykinin-induced shock and *E. coli* peritonitis have shown less endotoxemia and enhanced survival in animals receiving intra-intestinal instillation of non-absorbable antibiotics (Cuevas and Fine 1971).

2.4 Human studies

Studies in man on the protective effects of polymyxins in endotoxemia have been few and poorly controlled. Corrigan et al (1974) gave human volunteers polymyxin B (0.07 mg/kg) mixed with the second dose of typhoid-paratyphoid vaccine. The neutrophilic response seen in controls after vaccination was delayed by 4 hours in the polymyxin-treated individuals. However, immunogenicity of the combination was not studied.

In a few uncontrolled studies polymyxin B has been used to treat patients with circulating endotoxin activity associated with liver cirrhosis (Liehr et al 1975; Wilkinson et al 1976). Endotoxemia was eradicated in 4 of 7 cirrhotic patients given 1.1 mg/kg/day of polymyxin B for 3 days. Five of 9 patients improved clinically with elevation of fibrinogen levels and platelet counts; renal function improved in 4 patients. However, 2 patients developed endotoxemia while on polymyxin therapy. Scevola et al (1979) found that oral administration of paromomycin and lactulose abolished endotoxemia in 18 or 19 patients with liver disease with concomitant reduction in ammonia levels. These results are more likely due to decreased production of endotoxin in the gut rather than to a specific antiendotoxin effect of paromomycin.

McCabe (1975) compared retrospectively the fate of 288 patients with gram-negative bacteremia treated with polymyxin B or colistin or other appropriate antibiotics. In all cases the causative organisms were sensitive to the administered antimicrobial agent. The incidence of shock and death was significantly more frequent in patients receiving polymyxins than in those treated with other antibiotics (36% *vs* 17%, $p < 0.01$). However, no information is available on whether the polymyxin group of patients had more infections due to *Pseudomonas aeruginosa*, which seems likely. Nevertheless, these data suggest that the anti-endotoxic effect of polymyxins has no major role in the therapy of patients with gram-negative infections.

Thus, the use of polymyxins in clinical situations seems to be fraught with problems. Firstly, it is not clear whether all or any of the manifestations of gram-negative bacteria and shock are mediated or caused by endotoxins (Kass et al 1973; Stumacher et al 1973). Secondly, to be effective in animal models, polymyxins need to be administered very close to endotoxin challenge, a situation one is unlikely to encounter clinically. Thirdly, relatively large doses of the polymyxins, which in themselves are potentially toxic, are required to neutralize endotoxin. It is doubtful, therefore, that further studies will establish a clear role for these drugs in the therapy of endotoxemia. On the other hand, the polymyxins have contributed significantly to our understanding of endotoxin action and research in this area will undoubtedly continue for years to come.

3.   ENDOTOXIN RELEASE BY ANTIMICROBIAL AGENTS

The mortality rate due to gram-negative infections, in particular those associated with shock, is still high in spite of progress in antimicrobial therapy. It has long been speculated that endotoxin release following lysis of organisms may contribute to this. Surprisingly, little effort has been directed towards this question. Most work has focused on the Jarisch-Herxheimer reaction after therapy of spirochetal infections. Gelfand et al (1976) demonstrated circulating endotoxin by Limulus lysate assay 4–8 hours after penicillin therapy of two patients with secondary syphilis. In another study, 11 of 15 patients with *Borrelia recurrentis* infection had positive Limulus tests prior to therapy and 13 of 15 had positive tests 2 hours after treatment. At 7 days after treatment only 4 patients had a positive Limulus lysate assay (Galloway et al 1977). Serum from patients with the Jarisch-Herxheimer reaction has also been shown to cause fever in rabbits (Bryceson et al 1972; Bryceson 1976). Serum obtained the day after therapy when the patients were asymptomatic induced little or no fever in the rabbits.

While these data are interesting, they may not necessarily reflect the situation in gram-negative bacterial infections. The biological activity of spirochetal lipopolysaccharides may be different from that of LPS from gram-negative organisms. Indeed, in some studies spirochetal lipopolysaccharides have been shown to contain no pyrogenic activity or to be much less potent on a weight basis than bacterial products (Johnson 1976; Mergenhagen et al 1961). Jarisch-Herxheimer-like reactions after therapy of gram-negative infections are poorly documented. However, in a now classic study on therapy of severe *Brucella melitensis* infections in Mexico City, twelve of 24 patients treated with chlortetracycline developed an abrupt rise in temperature 8–12 hours after the first dose of drug occasionally accompanied by shock (Spink et al 1948). This reaction was seen primarily in patients who received a dose of 0.5 g 4 times a day. When an initial daily dose of 0.1 g was used and increased gradually to 2 g per day by the fourth day, fewer of these reactions were observed. Patients with *Brucella suis* and *Brucella abortus* infections, which tend to be less severe than *B. melitensis* infections, did not experience this reaction even at 2–6 g of chlortetracycline per day.

This phenomenon has not been well described since and may not be clinically apparent except in very ill patients with high-grade bacteremia where marked lysis of organisms could occur soon after onset of therapy. In an attempt to reproduce this in an animal model, Shenep and Mogan (1983) examined the effect of gentamicin therapy of *E. coli* bacteremia in rabbits on circulating endotoxin activity. A 10–2000-fold rise in free endotoxin levels was observed in rabbit sera after 6 hours of therapy. Endotoxin levels in untreated animals rose 20–400-fold during the same time; however, the increase in endotoxin activity was generally less than the corresponding increase in the number of bacteria in the blood. In contrast, the magnitude of bacteremia in most treated animals decreased during treatment with gentamicin. Maximum rise in endotoxin levels occurred after most bacteria in the bloodstream were eliminated, suggesting an additional source of endotoxin from bacteria at tissue sites. Preliminary data from the same investigators indicated that third-generation cephalosporins caused even greater increases in endotoxemia than gentamicin. Clearly, further investigations into this area of research are indicated.

4. EFFECTS OF ENDOTOXIN ON ANTIMICROBIAL PHARMACOKINETICS

The effect of endotoxin or even shock on the absorption, distribution and elimination of antibiotics has been poorly studied. However, investigations with other drugs in patients with shock, hypotension and cardiac failure (Benet et al 1976; Benowitz et al 1974; Feely et al 1982; Ritz et al 1973; Thomson et al 1973; Wilkinson 1976) suggest that there would be slower gastrointestinal and intramuscular absorption, a smaller volume of distribution, decreased hepatic and renal clearances and a longer serum half-life. These alterations in drug pharmacokinetics should result in higher serum levels, but lower bile and urine concentrations. Studies in rats have demonstrated higher serum concentrations of gentamicin, tobramycin, amikacin, netilmicin, ampicillin and cephalothin when administered as a single dose 3 hours after endotoxin challenge (Beauchamp et al 1982; Bergeron and Bergeron 1983). In contrast, urinary concentrations of these drugs were reduced 25–71%.

Tissue concentrations may also be increased. The higher serum levels would provide a larger concentration gradient from blood to interstitial fluid. Craig et al (1976) have demonstrated that gram-negative infections increase serum free fatty acids which in turn displace drugs from albumin binding sites. Decreased protein binding would also be expected to enhance penetration of antimicrobials into tissues. A slower capillary flow as in shock would provide more time for an antibiotic to pass through capillary pores into the interstitial fluid. Larsson et al (1983), using an in vitro capillary diffusion model, have demonstrated increased diffusion of ampicillin and dicloxacillin at reduced flow rates. The only data on the effect of endotoxin on antimicrobial tissue levels are from the previously mentioned studies in rats (Beauchamp et al 1982; Bergeron and Bergeron 1983). These investigators observed that prior challenge with endotoxin increased the kidney cortex and medulla concentrations of all four aminoglycoside and two β-lactam antibiotics studied. These observations may have both toxicologic and therapeutic importance, but any extrapolation to man is obviously premature at this stage. This is another area of research in which more information is needed.

5. ANTIMICROBIAL THERAPY IN GRAM-NEGATIVE BACTEREMIA AND SHOCK

There are a large number of studies evaluating the efficacy of different antimicrobial agents in bacteremic animal model infections with gram-negative bacteria. However, only a few studies involving rats, dogs and monkeys have employed antibiotics in the presence of shock (Balis et al 1979; Hinshaw et al 1979, 1980; Pitcairn et al 1975; Saslaw et al 1972; Short et al 1981). Although mortality in the presence of shock was reduced by antibiotic therapy, it was not eliminated. None of these studies used maximum antimicrobial therapy; combinations were rarely used, and the frequency of dosing would allow long periods of subinhibitory concentrations. In fact, most of these studies were designed to evaluate other modalities such as steroids alone or in combination with antibiotics.

A discussion of the role of antimicrobial therapy in gram-negative shock is inevitably based on the experience with therapy of gram-negative bacteremia in general. In fact, endotoxin shock probably represents just a more severe form of the same disease process. An assessment of the efficacy of antimicrobial therapy in gram-negative bacteremia is, of course, complicated by variations in patients and underlying diseases and the variable use of other therapeutic modalities. Nevertheless, the discussion that follows will evaluate:

a.   the efficacy of appropriate versus inappropriate therapy in gram-negative bacteremia with and without shock;

b.   the efficacy of antimicrobial therapy in preventing shock;

c.   the efficacy of single versus combination antimicrobial therapy;

d.   the importance of synergistic combinations, peak serum levels and the serum bactericidal titer.

5.1   Appropriate versus inappropriate therapy

Eleven major studies on gram-negative bacteremia were published during the past 20 years that confirm the benefit of appropriately selected antibiotics. The results and statistical significance with each study are listed in Table 2. Mortality rates ranged from 6–33% in the appropriately treated patients and 30–53% in the nonneutropenic patients and 70–100% in the neutropenic patients treated with inappropriate antibiotics. Patients with rapidly fatal underlying disease treated before 1970 did not significantly benefit from appropriate antibiotic therapy. For example, if one pools the four studies based on therapy before 1970 (Bryant et al 1971; Dupont and Spink 1969; Freid and Vosti 1968; McCabe and Jackson 1962), mortality in the appropriate and inappropriate treatment groups was 82 and 86%, respectively. The use of appropriate antimicrobial therapy did significantly lower mortality in patients with ultimately fatal and nonfatal underlying diseases. With the advent of more aggressive antimicrobial therapy and better support in the 1970s appropriate antibiotics have now become of significant benefit even for patients with rapidly fatal underlying diseases (Bryan et al 1983; EORTC 1978; Kreger et al 1980; Love et al 1980).

A recent study by Bryan et al (1983) suggested that choice of antimicrobial therapy on the calendar day when blood cultures first became positive was not related to eventual clinical outcome. Mortality was 36% in the appropriately treated patients and 39% in those who received incorrect or no therapy. It was only when these investigators included appropriate therapy started after the first calendar day that a beneficial effect of antimicrobial therapy could be demonstrated. In fact, the timing of appropriate therapy was of no importance in this study, since survival was independent of whether

248

TABLE 2　*Effect of appropriate versus inappropriate antimicrobial therapy on mortality of gram-negative bacteremia*

Appropriate treatment			Inappropriate treatment					
No. of patients	Mortality		No. of patients	Mortality		χ^2	p	Reference
	No.	Rate(%)		No.	Rate(%)			
81	18	22	31	15	48	7.4	<0.01	McCabe and Jackson 1962
212	68	32	58	28	48	5.2	<0.025	Freid and Vosti 1968
181	59	33	68	36	53	8.7	<0.001	Dupont and Spink 1969
158	42	27	60	28	47	8.0	<0.005	Bryant et al 1971
95	6	6	34	16	47	29.4	<0.0005	Myerowitz et al 1971
63	12	19	14	9	64	11.8	<0.001	Setia and Gross 1977
352	80	23*	92	37	40*	11.5	<0.001	Anderson et al 1978
70	21	30	6	6	100	9.3	<0.005	EORTC 1978
431	84	19	172	64	37	20.8	<0.0005	Kreger et al 1980
60	14	23*	7	5	71*	5.0	<0.05	Love et al 1980
1017	175	17	169	51	30	15.8	<0.001	Bryan et al 1983

* Represents failure of response rather than mortality

appropriate antibiotics were started 1, 2, or even 3 or more days after blood cultures first became positive.

The majority of patients included in the studies listed in Table 2 did not have clinical shock. In the seven studies in which this information was given, the frequency of shock at the initiation of antimicrobial therapy ranged from 13–44%. Only three of these studies have separately analyzed the effect of appropriate versus inappropriate antimicrobial therapy on mortality in these patients. These data are included in Table 3 along with the results of two early studies performed only in patients with shock. Mortality was clearly higher in these patients than in all patients with gram-negative bacteremia (Table 2). It ranged from 41–78% in the appropriately treated patients and 50–100% in those who were incorrectly treated. Appropriate antimicrobial therapy was not as effective in preventing death as in other patients; in fact, in two of the older studies no statistical significance could be demonstrated between the two treatment groups. Only Kreger et al (1980) had enough patients to examine survival by underlying disease categories. They observed a lower mortality rate in all three disease categories, but the difference was not statistically significant among the patients with ultimately fatal underlying disease.

5.2 Antimicrobial therapy in prevention of shock

The previously mentioned study by Bryan et al (1983) which suggested that the timing of appropriate antimicrobial therapy was not important in determining outcome should not lead to therapeutic nihilism concerning initial empiric therapy. Kreger et al (1980) have collected data indicating that appropriate antimicrobial therapy early in the course of bacteremia may prevent the subsequent occurrence of shock. They were able to retrospectively analyze the occurrence of shock after initiation of appropriate or inappropriate antimicrobial therapy in 498 patients. Shock subsequently developed in 21% of patients receiving appropriate antibiotics and in 49% of those receiving incorrect therapy ($\chi^2 = 41.3$; $p < 0.001$). This at least two-fold reduction in the frequency of shock was observed in patients with rapidly fatal, ultimately fatal and nonfatal underlying diseases.

5.3 Single agent versus combination therapy

There are no studies that have directly compared single agent versus combination therapy in patients with gram-negative bacteremic shock. However,

TABLE 3 *Effect of appropriate and inappropriate antimicrobial therapy on mortality in gram-negative bacteremic shock*

Appropriate therapy			Inappropriate treatment					
No. of patients	Mortality		No. of patients	Mortality				
	No.	Rate(%)		No.	Rate(%)	χ^2	p	Reference
107	83	78	23	21	91	2.2	N.S.	Weil et al 1964
44	22	50	16	16	100	9.1	<0.005	Bryant et al 1971
35	18	51	8	4	50	0.1	N.S.	Christy 1971
182	75	41	88	52	59	7.6	<0.01	Kreger et al 1980
151	86	57	25	21	84	6.6	<0.025	Bryan et al 1983

N.S. = not significant

several studies have appeared in the literature over the past ten years that have addressed this question. Most of these reports have been limited to strictly immunocompromized patients, and obviously such data are probably not applicable to the normal host. Only two of the studies have been prospective comparisons of single versus combination drug therapy (Bodey et al 1976; Klastersky et al 1973). Two others represent retrospective comparisons of therapy with a single drug to that with multiple drugs (Klastersky et al 1974; Kreger et al 1980). In the remaining studies combination therapy was used, but efficacy was compared in patients in which the infecting organisms were susceptible to only one or to both of the agents in the combination. The results of these investigations as they relate to gram-negative bacteremia are included in Table 4.

Although the available evidence seems to suggest a trend in favor of combination therapy in both prospective and retrospective trials, only two studies showed a significant difference between single and combination drug therapy. The study by Klastersky et al (1974) demonstrated a better response for combination therapy. However, these data do include a few gram-positive bacteremias, and it is unknown if the difference would still be significant if those cases were eliminated from analysis. The study by Anderson et al (1978) demonstrated a better response with single drug therapy. However, this was not a randomized prospective study and a much larger percentage of bacteremias arising from the urinary tract were treated with a single drug. Thus, no single study has proven the superiority of combination over single drug agents for gram-negative bacteremia.

5.4 Importance of synergy, serum levels and serum bactericidal activity

Three of the nine studies on single versus combination antimicrobial therapy in gram-negative bacteremia have also examined the importance of synergistic combinations on outcome. The results of these studies plus one additional trial are included in Table 5. The synergistic combinations which consisted primarily of an aminoglycoside and β-lactam antibiotic resulted in significantly better outcome in three of the four studies. The one trial not reaching statistical significance demonstrated the same trend but did not include a sufficient number of patients. The study by Anderson et al (1978) also examined various subgroups of patients. Significant benefit of synergistic combinations was observed in neutropenic patients, those with rapidly fatal underlying disease, patients in shock and those with *P. aeruginosa* infections. A

252

TABLE 4 *Effect of single agent versus combination therapy on response in gram-negative bacteremia*

Single agent therapy			Combination therapy					
No. of patients	Responses		No. of patients	Responses		χ^2	p	Reference
	No.	Rate(%)		No.	Rate(%)			
7	3	43	5	4	80	0.5	N.S.	Klastersky et al 1973
16	6	38	46	35	76	6.3	<0.025	Klastersky et al 1974
12	7	58	16	13	81	0.8	N.S.	Bodey et al 1976
24	14	58	22	13	59	<0.1	N.S.	Klastersky et al 1977
7	2	29	31	23	74	3.4	N.S.	Lau et al 1977
179	147	82	173	124	72	5.4	<0.025	Anderson et al 1978
19*	8	42	54*	31	57	0.8	N.S.	Anderson et al 1978
32	22	69	38	27	71	<0.1	N.S.	EORTC 1978
27	18	67	33	28	85	1.8	N.S.	Love et al 1980
231	182	79	260	203	78	<0.1	N.S.	Kreger et al 1980
14*	7	50	22*	17	77	1.8	N.S.	Kreger et al 1980

* Only patients with rapidly fatal underlying disease.
N.S. = not significant.

TABLE 5 *Effect of synergistic antimicrobial combinations on response in gram-negative bacteremia*

Synergism			No synergism					
No. of patients	Response		No. of patients	Response		χ^2	p	Reference
	No.	Rate(%)		No.	Rate(%)			
20	17	85	10	7	41	5.9	<0.05	Klastersky et al 1972
24	18	75	22	9	41	4.2	<0.05	Klastersky et al 1977
22	18	82	9	5	56	1.1	N.S.	Lau et al 1977
83	66	80	90	58	64	4.8	<0.05	Anderson et al 1978
26*	19	73	31*	13	42	5.6	<0.025	Anderson et al 1978

* Subgroup of patients with rapidly fatal underlying disease.
N.S. = not significant.

similar trend was seen in patients with ultimately fatal underlying disease and those with *Klebsiella pneumoniae* infections, but these results did not quite reach statistical significance.

The relationship of serum concentrations to the minimum inhibitory or bactericidal concentration of drug for the infecting organism has been suspected, for a long time, to have importance in determining outcome. A few studies have demonstrated that low peak serum levels of the aminoglycosides are associated with a higher failure rate in gram-negative infections in neutropenic patients (Keating et al 1979; Noone et al 1974; Riff and Jackson 1971). For example, in the study by Keating et al (1979), the clinical response in patients with optimal peak serum aminoglycoside levels was 76% compared with a response rate of 61% in those with low peak levels. This difference was significant at the 5% level. Anderson et al (1976) have also demonstrated that about half of their episodes of 'breakthrough' gram-negative bacteremia were associated with subinhibitory antibiotic levels.

A better way to relate serum concentrations to the minimum bactericidal concentration (MBC) is to determine the serum bactericidal activity against the infecting organism. Klastersky et al (1974) observed that the mean peak serum bactericidal titer was 1:16 in septicemias treated successfully, while the mean titer was only 1:4 in those that failed therapy. The effect of peak serum bactericidal activity on efficacy of antibiotic therapy in gram-negative bacteremia from two subsequent studies are shown in Table 6. Response was higher in patients in peak serum bactericidal titers of 1:8 or greater but reached statistical significance in only one of the two studies. A similar differ-

TABLE 6   *Effect of peak serum bactericidal activity (SBA) on response in gram-negative bacteremia*

Peak SBA	No. of patients	Response		χ^2	p	Reference
		No.	Rate (%)			
<1:4	17	8	47	3.3	<0.1	Klastersky et al 1977
≥1:8	24	19	79			
<1:4	10	1	10	35.5	<0.0005	Klastersky 1983
≥1:8	69	64	93			

ence was observed in both neutropenic and non-neutropenic patients (Klastersky 1983).

6. SUMMARY

The clinical studies examined in this review clearly demonstrate that appropriate antibiotics are beneficial in the patient with gram-negative bacteremia and shock. Not only do they reduce mortality but they also reduce the occurrence of shock. Evidence from both prospective and retrospective studies tend to favor the use of synergistic over nonsynergistic drug combinations or single drug therapy in immunocompromised patients. However, there is no convincing evidence that combinations of antimicrobials are better than a single active drug in patients without severe underlying disease. Limited data on serum drug levels and serum bactericidal activity suggest that the more susceptible the organism is in relation to the achieved serum levels, the greater the likelihood of clinical response. With the advent of new β-lactam antibiotics with enhanced antimicrobial activity against gram-negative bacteria, it is possible that the use of these drugs alone may be as effective as the older antibiotics when used in combination. This needs to be evaluated in controlled prospective clinical trials with specific focus on individual organisms and underlying diseases. Other areas requiring further research include the role of different antimicrobial dosage regimes, the influence of endotoxin and shock on antimicrobial pharmacokinetics, and the potential effects of antimicrobials on bacterial lysis with subsequent release of endotoxin. The role of antibiotics in endotoxin shock will probably continue to be an evolving area of research for many years to come.

REFERENCES

Anderson ET, Young LS, Hewitt WL (1976) Simultaneous antibiotic levels in 'breakthrough' gram-negative rod bacteremia. *Am. J. Med. 61*, 493-497.
Anderson ET, Young LS, Hewitt WL (1978) Antimicrobial synergism in the therapy of gram-negative rod bacteremia. *Chemotherapy 24, 45-54.*
Axelrod BJ, Shands JW (1977) Commitment to deoxyribonucleic acid synthesis and the cell cycle in endotoxin-stimulated murine spleen cells. *Infect. Immun. 17*, 344-349.

Bader J, Teuber M (1973) Action of polymyxin B on bacterial membranes. I. Binding to the O-antigenic lipopolysaccharide of *Salmonella typhimurium*. *Z. Naturforsch.* *28c*, 422-430.

Balis JU, Paterson JF, Shelley SA, Larson CH, Fareed J, Gerber LI (1979) Glucocorticoid and antibiotic effects on hepatic microcirculation and associated host responses in lethal gram-negative bacteremia. *Lab. Invest. 40*, 55-65.

Bannatyne RM, Cheung R (1977) Protective effect of polymyxin B sulfate in experimental meningococcal infection in mice. *Can.J. Microbiol. 23*, 1526-1528.

Bannatyne RM, Cheung R (1979) Protective effect of polymyxin B sulfate in experimental enterobacterial infection in mice. *Can. J. Microbiol. 25*, 995-998.

Bannatyne RM, Harnett NM, Lee K, Biggar WE (1977) Inhibition of the biologic effects of endotoxin on neutrophils by polymyxin B sulfate. *J. Infect. Dis. 136*, 469-474.

Barnett JA, Sanford JP (1969) Bacterial shock. *J.Am. Med. Assoc. 209*, 1514-1517.

Beauchamp D, Bergeron Y, Bergeron MG (1982) Influence of endotoxin on the intrarenal distribution of gentamicin. In: *Program and Abstracts of the 22nd Interscience Conference on Antimicrobial Agents and Chemotherapy*, Abstract 45. American Society for Microbiology, Washington, D.C.

Benet LZ, Greither A, Meister W (1976) Gastrointestinal absorption of drugs in patients with cardiac failure. In: Benet LZ (Ed), *The Effect of Disease States on Drug Pharmacokinetics*, pp 35-50. American Pharmaceutical Association and Academy of Pharmaceutical Sciences, Washington, D.C.

Benowitz N, Forsyth RP, Melman KL Rowland M (1974) Lidocaine disposition kinetics in monkey and man. II. Effects of hemorrhage and sympathomimetic drug administration. *Clin. Pharmacol. Ther. 16*, 99-109.

Bergeron MG, Bergeron Y (1983) Influence of endotoxin on the intrarenal distribution of antibiotics. In: *Program and Abstracts of the 23rd Interscience Conference on Antimicrobial Agents and Chemotherapy*, Abstract 596. American Society for Microbiology, Washington, D.C.

Bergstein JM, Michael AF Jr (1973) Effect of polymyxin B and endotoxin on renal cortical fibrinolysis. *Proc. Soc. Exp. Biol. Med. 144*, 369-372.

Bodey GP, Feld R, Burgess MA (1976) β-Lactam antibiotics alone or in combination with gentamicin for therapy of gram-negative bacillary infections in neutropenic patients. *Am. J. Med. Sci. 271*, 179-186.

Braude AI, Carey FJ, Zalesky M (1955) Studies with radioactive endotoxin. II. Correlation of physiologic effects with distribution of radioactivity in rabbits injected with lethal dose of E. coli endotoxin labelled with radioactive sodium chromate. *J. Clin. Invest. 34*, 858-866.

Bryan CS, Reynolds KL, Brenner FR (1983) Analysis of 1186 episodes of gram-negative bacteremia in non-university hospitals: the effects of antimicrobial therapy. *Rev. Infect. Dis. 5*, 629-638.

Bryant RE, Hood AF, Hood CE, Koenig MG (1971) Factors affecting mortality of gram-negative rod bacteremia. *Arch. Intern. Med. 127*, 120-128.

Bryceson ADM (1976) Clinical pathology of the Jarisch-Herxheimer reaction. *J. Infect. Dis. 133*, 696-704.

Bryceson ADM, Cooper KE, Warrell DA, Perina PL, Parry EHO (1972) Studies of the mechanism of the Jarisch-Herxheimer reaction in louse-borne relapsing fever: evidence for the presence of circulating Borrelia endotoxin. *Clin. Sci. 43*, 343-354.

Butler T, Möller G (1977) Mitogenic response of mouse spleen cells and gelation of limulus lysate by lipopolysaccharide of Yersinia pestis and evidence for neutralization of the lipopolysaccharide by polymyxin B. *Infect. Immun. 18*, 400-404.

Christy JH (1971) Treatment of gram-negative shock. *Am. J. Med. 50*, 77-88.

Cooperstock MS (1974) Inactivation of endotoxin by polymyxin B. *Antimicrob. Agents Chemother. 6*, 422-425.

Cooperstock MS, Riegle L (1981) Polymyxin B inactivation of lipopolysaccharide in vaccines of gram-negative bacteria. *Infect. Immun. 33*, 315-318.

Corrigan JJ Jr, Bell BM (1971a) Endotoxin-induced intravascular coagulation: prevention with polymyxin B sulfate. *J. Lab. Clin. Med. 77*, 802-810.

Corrigan JJ Jr, Bell BM (1971b) Comparison between the polymyxins and gentamicin in preventing endotoxin-induced intravascular coagulation and leukopenia. *Infect. Immun. 4*, 563-566.

Corrigan JJ Jr, Kiernat JF (1979) Effect of polymyxin B sulfate on endotoxin activity in gram-negative septicemia model. *Pediatr. Res. 13*, 48-51.

Corrigan JJ Jr, Sieber OF Jr, Ratajczak H, Bennett BB (1974) Modification of human neutrophil response to endotoxin with polymyxin B sulfate. *J. Infect. Dis. 130*, 384-387.

Craig WA, Turner JH, Kunin CM (1974) Prevention of the generalized Shwartzman reaction and endotoxin lethality by polymyxin B localized in tissues. *Infect. Immun. 10*, 287-292.

Craig WA, Evenson MA, Ramgopal V (1976) The effect of uremia, cardiopulmonary bypass and bacterial infection on serum protein binding. In: Benet LZ (Ed), *The Effect of Disease States a Drug Pharmacokinetics*, pp 125-136. American Pharmaceutical Association and Academy of Pharmaceutical Sciences, Washington, D.C.

Cuevas P, Fine J (1971) Demonstration of a lethal endotoxinemia in experimental occlusion of the superior mesenteric artery. *Surg. Gynecol. Obstet. 133*, 81-83.

Davis TD, Iannetta A, Wedgwood RJ (1971) Activity of colistin against *Pseudomonas aeruginosa*. Inhibition by calcium. *J. Infect. Dis. 124*, 610-612.

DuPont HL, Spink WW (1969) Infections due to gram-negative organisms: an analysis of 860 patients with bacteremia at the University of Minnesota Medical Center, 1958-1966. *Medicine 48*, 307-332.

EORTC International Antimicrobial Therapy Project Group (1978) Three antibiotic regimens in the treatment of infection in febrile granulocytopenic patients with cancer. *J. Infect. Dis. 137*, 14-29.

Escartín P, Rodriguez-Montes JA, Cuervas-Mons V, Rossi I, Alvarez-Cienfuegos J, Maganto P, Castillo-Olivares JL (1982) Effect of colistin on reduction of biliary flow induced by endotoxin of E. coli. *Dig. Dis. Sci. 27*, 875-879.

Feely J, Wade D, McAllister CB, Wilkinson GR, Robertson D (1982) Effect of hypotension on liver blood flow and lidocaine disposition. *N. Engl. J. Med. 307*, 866-869.

Freid MA, Vosti KL (1968) The importance of underlying disease in patients with gram-negative bacteremia. *Arch. Intern. Med. 121*, 418-423.

From AHL, Fong JSC, Good RA (1979) Polymyxin B sulfate modification of bacterial endotoxin: effects on the development of endotoxin shock in dogs. *Infect. Immun. 23*, 660-664.

Galloway RE, Levin J, Butler T, Naff GB, Goldsmith GH, Saito H, Awoke S, Wallace CK (1977) Activation of protein mediators of inflammation and evidence for endotoxemia in *Borrelia recurrentis* infection. *Am. J. Med. 63*, 933-938.

Gans H, Matsumoto K (1974) Are enteric endotoxins able to escape from the intestine? *Proc. Soc. Exp. Biol. Med. 147*, 736-739.

Gelfand JA, Elin RJ, Berry FW, Frank MM (1976) Endotoxemia associated with the Jarisch-Herxheimer reaction. *N. Engl. J. Med. 295*, 211-213.

Gleckman R, Esposito A (1981) Gram-negative bacteremia shock: pathophysiology clinical features and treatment. *South. Med. J. 74*, 335-341.

Gruninger RP, Howe CW (1969) Temporal protection of chick embryo against endotoxin by isoproterenol, polymyxin B sulfate and hydrocortisone. *Proc. Soc. Exp. Biol. Med. 130*, 1202-1205.

Hassen A (1973) Gram-negative shock. *Med. Clin. North Am. 57*, 1403-1415.

Hinshaw LB, Beller BK, Archer LT, Flournoy DJ, White GL, Phillips RW (1979) Recovery from lethal *Escherichia coli* shock in dogs. *Surg. Gynecol. Obstet. 149*, 545-553.

Hinshaw LB, Archer LT, Beller-Todd BK, Coalson JJ, Flournoy DJ, Passey R, Benjamin B, White GL (1980) Survival of primates in LD$_{100}$ septic shock following steroid/antibiotic therapy. *J. Surg. Res. 28*, 151-170.

Horton J, Pankey GA (1982) Polymyxin B, colistin, and sodium colistimethate. *Med. Clin. North Am. 66*, 135-142.

Hughes B, Madan BR, Parratt JR (1981) Polymyxin B sulphate protects cats against the haemodynamic and metabolic effects of E. coli endotoxin. *Br. J. Pharmacol. 74*, 701-707.

Inada K (1980) Complement activating property of the protein-rich endotoxin (OEP) of *Pseudomonas aeruginosa*. II. Complement activating property of the lipopolysaccharide portion and the inhibition by polymyxin B. *Jpn. J. Exp. Med. 50*, 107-115.

Ingoldby CJH (1980) The value of polymyxin B in endotoxinaemia due to experimental obstructive jaundice and mesenteric ischaemia. *Br. J. Surg. 67*, 565-567.

Issekutz AC, Biggar WD (1978) Effect of methylprednisolone and polymyxin B sulfate on endotoxin-induced inhibition of human neutrophil chemotaxis. *J. Lab. Clin. Med. 92*, 873-882.

Jacobs DM, Morrison DC (1975) Dissociation between mitogenicity and immunogenicity of TNP-lipopolysaccharide, a T-independent antigen. *J. Exp. Med. 141*, 1453-1458.

Jacobs DM, Morrison DC (1977) Inhibition of mitogenic response to lipopolysaccharide (LPS) in mouse spleen cells by polymyxin B. *J. Immunol. 118*, 21-27.

Johnson RC (1976) Comparative spirochete physiology and cellular composition. In: Johnson RC (Ed.), *The Biology of Parasitic Spirochetes*, pp 39-48. Academic Press, New York.

Kass EH, Porter P, McGill MW, Vivaldi E (1973) Clinical and experimental observations on the significance of endotoxinemia. *J. Infect. Dis. 128*, S299-S302.

Keating MJ, Bodey GP, Valdivieso M, Rodriguez V (1979) A randomized comparative trial of three aminoglycosides-comparison of continuous infusions of gentamicin, amikacin and sisomicin combined with carbenicillin in the treatment of infections in neutropenic patients with malignancies. *Medicine 58*, 159-170.

Klastersky J (1983) Clinical significance of the serum bactericidal activity. In: Spitzy KH, Karrer K (Eds), *Proceedings of the 13th International Congress of Chemotherapy, Part 4*, pp 20-24. Verlag Egermann, Vienna, Austria.

Klastersky J, Cappel R, Daneau D (1972) Clinical significance of in vitro synergism between antibiotics in gram-negative infections. *Antimicrob. Agents Chemother. 2*, 470-475.

Klastersky J, Cappel R, Daneau D (1973) Therapy with carbenicillin and gentamicin for patients with cancer and severe infections caused by gram-negative rods. *Cancer 31*, 331-336.

Klastersky J, Daneau D, Swings G, Weerts D (1974) Antibacterial activity in serum and urine as a therapeutic guide in bacterial infections. *J. Infect. Dis. 129*, 187-193.

Klastersky J, Meunier-Carpentier F, Prevost JM (1977) Significance of antimicrobial synergism for the outcome of gram-negative sepsis. *Am. J. Med. Sci. 273*, 157-167.

Kocsar LT, Bertok L, Varterezz V (1969) Effect of bile acids on the intestinal absorption of endotoxin in rats. *J. Bacteriol. 100*, 220-223.

Kreger BE, Craven DE, McCabe WR (1980) Gram-negative bacteremia. IV. Re-evaluation of clinical features and treatment in 612 patients. *Am. J. Med. 68*, 344-355.

Landesman SH, Gorbach SL (1978) Gram-negative sepsis and shock. *Orthop. Clin. North. Am. 9*, 611-625.

Larsson SE, Larsson JE, Holm SE (1983) The importance of capillary flow rate for the penetration of antibiotics. In: Spitzy KH, Karrer K (Eds), *Proceedings of the*

13th International Congress of Chemotherapy, Part 85, pp 61-65. Verlag Egermann, Vienna, Austria.

Lau WK, Young LS, Black PE, Winston DJ, Linné SR, Weinstein RJ, Hewitt WL (1977) Comparative efficacy and toxicity of amikacin/carbenicillin versus gentamicin/carbenicillin in leukopenic patients. A randomized perspective trial. *Am. J. Med. 62*, 959-966.

Liehr H, Grün M, Brunswig D, Sautter T (1975) Endotoxinaemia in liver cirrhosis: treatment with polymyxin B. *Lancet 1*, 810-811.

Lopes J, Inniss WE (1969) Electronmicroscopy of effect of polymyxin on Escherichia coli lipopolysaccharide. *J. Bacteriol. 100*, 1128-1130.

Love LJ, Schimpff SC, Schiffer CA, Wiernik PM (1980) Improved prognosis for granulocytopenic patients with gram-negative bacteremia. *Am. J. Med. 68*, 643-648.

McCabe WR (1973) *Gram-negative bacteremia*. DM Disease-a-Month (December), Year Book Medical Publishers, Chicago.

McCabe WR (1975) Antibiotics and endotoxin shock. *Bull. NY Acad. Med. 51*, 1084-1094.

McCabe WR, Jackson GG (1962) Gram-negative bacteremia. II. Clinical, laboratory and therapeutic observations. *Arch. Intern. Med. 110*, 856-864.

McCabe WR, Olans RN (1981) Shock in gram-negative bacteremia: predisposing factors, pathophysiology and treatment. In: Remington JS, Swartz MN (Eds), *Current Clinical Topics in Infectious Disease, Vol 2*, pp 121-150. McGraw-Hill, New York.

Mergenhagen SE, Hampp EG, Scherp HW (1961) Preparation and biological activities of endotoxins from oral bacteria. *J. Infect. Dis. 108*, 304.

Morrison DC, Jacobs DM (1976a) Binding of polymyxin B to the lipid A portion of bacterial lipopolysaccharides. *Immunochemistry 13*, 813-818.

Morrison DC, Jacobs DM (1976b) Inhibition of lipopolysaccharide-initiated activation of serum complement by polymyxin B. *Infect. Immun. 13*, 298-301.

Myerowitz RL, Medeiros AA, O'Brian TF (1971) Recent experience with bacillemia due to gram-negative organisms. *J. Infect. Dis. 124*, 239-246.

Neter E, Gorzynski EA, Westphal O, Lüderitz O (1958) The effects of antibiotics on enterobacterial lipopolysaccharides (endotoxins), hemagglutination and hemolysis. *J. Immunol. 80*, 66-72.

Newton BA (1956) The properties and mode of action of the polymyxins. *Bacteriol. Rev. 20*, 14-27.

Noone P, Parsons TMC, Pattison JR, Slack RCB, Garfield-Davies D, Hughes K (1974) Experience in monitoring gentamicin therapy during treatment of serious gram-negative sepsis. *Br. Med. J. 1*, 477-481.

Palmer JD, Rifkind D (1974) Neutralization of the hemodynamic effects of endotoxin by polymyxin B. *Surg. Gynecol. Obstet. 138*, 755-759.

Pitcairn M, Shuler J, Erve PR, Holtzman S, Shumer W (1975) Glucocorticoid and antibiotic effect on experimental gram-negative bacteremic shock. *Arch. Surg. 110*, 1012-1015.

Rietschel ET, Gottert H, Lüderitz O, Westphal O (1972) Nature and linkages of the fatty acids present in the lipid-A component of salmonella lipopolysaccharides. *Eur. J. Biochem. 28*, 166-173.

Riff LJ, Jackson GG (1971) Pharmacology of gentamicin in man. *J. Infect. Dis. 124*, 98S-101S.

Rifkind D (1967a) Prevention by polymyxin B of endotoxin lethality in mice. *J. Bacteriol. 93*, 1463-1464.

Rifkind D (1967b) Studies on the interaction between endotoxin and polymyxin B. *J. Infect. Dis. 117*, 433-438.

Rifkind D, Hill RB (1967) Neutralization of the Shwartzman reactions by polymyxin B. *J. Immunol. 99*, 564-569.

Rifkind D, Palmer JD (1966) Neutralization of endotoxin toxicity in chick embryos by antibiotics. *J. Bacteriol. 92*, 815-819.

Ritz R, Cavanilles J, Michaels S, Shubin H, Weil MH (1973) Disappearance of indocyanine green during circulatory shock. *Surg. Gynecol. Obstet. 136*, 57-62.

Saslaw S, Carlisle MN, Moheimani M (1972) Comparison of tobramycin, gentamicin, colistin and carbenicillin in *Pseudomonas* sepsis in monkeys. *Antimicrob. Agents Chemother. 2*, 164-172.

Scevola D, Megliulo E, Trpin L, Benzi-Cipelli R, Bernardi R (1979) Control of endotoxinemia in liver disease by lactulose and paromomycin. *Boll. Ist. Sieroter. Milan. 58*, 242-247.

Setia U, Gross PA (1977) Bacteremia in a community hospital. Spectrum and mortality. *Arch. Intern. Med. 137*, 1698-1701.

Shenep JL, Mogan KA (1984) Kinetics of endotoxin release during antibiotic therapy for experimental gram-negative bacterial sepsis. *J. Infect. Dis. 150*, 380-388.

Short BL, Gardiner M, Walker RI, Jones SR, Fletcher JR (1981) Indomethacin improves survival in gram-negative sepsis. *Adv. Shock Res. 6*, 27-36.

Spink WW, Braude AL, Castaneda MR, Goytia RS (1948) Aureomycin therapy in human brucellosis due to Brucella melitensis. *J. Am. Med. Assoc. 138*, 1145-1148.

Stumacher RS, Kovnat MJ, McCabe WR (1973) Limitations of the usefulness of the limulus lysate assay for endotoxin. *N. Engl. J. Med. 288*, 1261-1264.

Sud IJ, Feingold DS (1975) Detection of agents that alter the bacterial cell surface. *Antimicrob. Agents Chemother. 8*, 34-37.

Thomson PD, Melman KL, Richardson JA, Cohn K, Steinbrunn W, Cudihee R, Rowland M (1973) Lidocaine pharmacokinetics in advanced heart failure, liver disease and renal failure in humans. *Ann. Intern. Med. 78*, 499-508.

Van Miert ASJPAM, Van Duin CThM (1977) The antipyretic effect of polymyxin B in endotoxin-induced fever. *Vet. Sci. Commun. 1*, 79-84.

Van Miert ASJPAM, Van Duin CThM (1978) Further studies on the antipyretic action of polymyxin B in pyrogen induced fever. *Arzneimittel. Forsch. 28*, 2246-2251.

Warren JR (1982) Polymyxin B suppresses the endotoxin inhibition of concanavalin A-mediated erythrocyte agglutination. *Infect. Immun. 35*, 594-599.

Weil MH, Shubin H, Biddle M (1964) Shock caused by gram-negative microorganisms. Analysis of 169 cases. *Ann. Intern. Med. 60*, 384-400.

Wilkinson GR (1976) Pharmacokinetics in disease states modifying body perfusion. In: Benet LZ (Ed), *The Effect of Disease States on Drug Pharmacokinetics*, pp 13-32. American Pharmaceutical Association and Academy of Pharmaceutical Sciences, Washington, D.C.

Wilkinson SP, Moodie H, Stamatakis JD, Kakkar VV, Williams R (1976) Endotoxaemia and renal failure in cirrhosis and obstructive jaundice. *Br. Med. J. 2*, 1415-1418.

Young LS (1982) Combination or single drug therapy for gram-negative sepsis. In: Remington JS, Swartz MN (Eds), *Current Clinical Topics in Infectious Disease, Vol 3*, pp 177-205. McGraw-Hill, New York.

Young LS, Màrtin WJ, Meyer RD, Weinstein RJ, Anderson ET (1977) Gram-negative rod bacteremia: microbiologic, immunologic and therapeutic considerations. *Ann. Intern. Med. 86*, 456-471.

Handbook of Endotoxin, Vol. 4: Clinical Aspects of Endotoxin Shock
R.A. Proctor, editor
© Elsevier Science Publishers B.V., 1986

CHAPTER 8

Role of glucocorticoids in the treatment of endotoxin shock

LOWELL S. YOUNG

1. INTRODUCTION

The appeal of corticosteroids for therapy of shock associated with gram-negative infections seems readily obvious when we consider their most prominent, immediate clinical effects. Corticosteroids have marked antipyretic and antiinflammatory properties. They make patients appear better, although this may be quite transient and a reduction in body temperature may give the impression of a response to the treatment of infection. Further, in some patients they induce euphoria and a feeling of well-being which further reinforce the subjective feeling of improvement – in both patient and observer. Suppression of the inflammatory response with reduction in body temperature have together provided many a clinician the basis for belief that corticosteroids are beneficial in the treatment of shock associated with endotoxin.

The latter point – that shock associated with gram-negative infections is 'endotoxin shock' – is by itself a contentious issue (Kreger et al 1980; Shine et al 1980). A vast number of animal studies, including some with nonhuman primates, have equated the administration of large doses of endotoxin to experimental subjects with the events associated with hypotension and decreased organ perfusion in humans secondary to infection with gram-negative rods. They have also served to justify the therapeutic use of corticosteroids in gram-negative infections. However, there are major reservations about the extrapolation of such studies to man. Man is the most sensitive of all organisms to the biological effects of endotoxin (Wolff 1973). Very small quantities of endotoxin have been used in the human studies to induce pyrexia, bradykinin generation, and other important clinical responses (Kimball et al 1972). Even the type of anesthesia may influence the induction of septic shock and susceptibility to the biological effects of endotoxin (McCabe 1974). The whole issue of whether endotoxin is a prime mediator in some of the events following gram-negative rod bacteremia has been widely debated (Kreger et al 1980; Young 1985).

2. REVIEWS OF EARLIER STUDIES

It is not surprising that the early efforts to use corticosteroids in man should have led to a highly variable picture of clinical efficacy that persists until this day. The important review article published by Weitzman and Berger in 1974 assessed 32 studies of corticosteroid therapy for bacterial infection. These studies appeared in the English language literature and addressed the issue of corticosteroid therapy on morbidity and mortality of human bacterial infection, excluding tuberculosis. All animal studies as well as in vitro studies were excluded from consideration. The analysis was conducted along lines described by Gifford and Feinstein (1969). Eight methodologic standards were chosen:
a.   prospective versus retrospective design;
b.   the presence or absence of concurrent controls;
c.   the use of random allocation of patients to treatment protocols;
d.   the use of a double-blind technique;
e.   a clear definition of diagnostic criteria;
f.   stratification of patients according to the clinical extent of disease;
g.   consideration of underlying disease;
h.   observations on any possible complications of corticosteroid therapy.

Each study was tested for adherence to these standards. Unfortunately, only 44% of 32 papers used a concurrent control population, only 41% were prospective experimental trials, and only 16% used a double-blind technique. Seventy-five percent of these studies did not allocate treatment in a random manner. Only 59% of the studies adequately described criteria for diagnosis of the illness treated. The presence of underlying disease was indicated in 12 studies and the clinical extent of disease was considered in only 14. Only 12 of the 32 studies evaluated potential complications of steroid administration. It seems likely that the duration of corticosteroid administration is an important determinant of both efficacy and toxicity. Nonetheless, in only 11 studies reviewed by Weitzman and Berger (1974) were patients treated in the experimental group for a specified number of days. Steroid dosage varied considerably within certain studies. In some cases, the actual dosage was expressed only in vague terms. Actually, only 12 studies or 38% specified a dosage regimen. Only 12 studies (38%) described the route of corticosteroid administration to patients.

Clearly, the effectiveness of corticosteroids may be related to the point at which they are initiated during the course of illness. If this point in time is not specifically indicated, the experimental group clearly becomes highly heterogeneous with respect to prognosis. Subsequent clinical events will thus not only reflect the effectiveness of corticosteroids (or their ineffectiveness) but also the previous therapy and the natural history of the disease. Only one study of the entire group initiated corticosteroid treatment at a single point in the illness. Two additional studies simply tabulated relevant data without controls.

Of all of the papers reviewed by Weitzman and Berger, an average of 1.7 standards per paper were satisfied. No studies fulfilled all five aspects of consistent design which the authors considered important. Nine papers confined their analyses to the effects of steroids for shock in patients with severe infection. Three additional papers concerned with severe infections contained a distinct subpopulation of patients in shock. These studies were, on the average, more remiss in adherence to methodologic criteria of clinical trial design. With only one exception, a study by Klastersky and colleages (1971), all papers on septic shock neglected to employ a double-blind study design and did not allocate therapy in a random manner. Weitzman and Berger concluded that of 32 reports, 22 considered steroids as a form of adjunctive therapy that favorably influenced the outcome of bacterial infection. Two concluded that steroids were harmful and 8 observed no difference between treatment and control groups. Included in the latter groups was the study reported by Klastersky and associates (1971). An attempt was made in this analysis to relate therapeutic conclusions with adherence to methodologic standards. Papers advocating steroid therapy were more delinquent in fulfilling satisfactory principles of clinical trial design than those that did not advocate steroids. Of those studies that treated patients in septic shock, there were more shortcomings with respect to the 8 methodologic criteria among studies advocating steroid therapy than in those that did not.

Weitzman and Berger concluded that most of the studies they reviewed failed to satisfy even half the 8 methodologic standards. One of the most important criteria for gram-negative bacteremia is the stratification of patients according to underlying disease, such as the fairly simple classification described by McCabe and Jackson (1962). Other crucial data that were not usually provided included specific steroid preparation used, its dosage, duration of therapy, and specification of the point in the course of illness when steroids were used. The authors of the review concluded that studies that did not satisfy their criteria were not necessarily unacceptable. Nonetheless, the historical perspective gained from this review suggested that inadequately designed studies perpetuate confusion and uncertainty and future studies with sufficient adherence to sound principles of clinical design were badly needed.

3. THE SCHUMER STUDY

In 1976, a paper written by a single author, William Schumer of the Westside Veterans Administration Hospital in Chicago, Illinois, claimed to overcome all of the previously published objections about the clinical design of trials evaluating use of corticosteroids in septic shock (Schumer 1976). This interesting study contained both a retrospective and a prospective component. Patients on entry to the study were given either methylprednisolone (30 mg/kg) or dexamethasone (3 mg/kg). These were obviously large doses, given at onset of arterial hypotension in a randomized prospective double-blind format with well-defined criteria for septic shock. At the discretion of the treating physician, the steroid bolus could be repeated in 4 hours. In the prospective component of the study, the results were remarkable in demonstrating a reduction in mortality from 38% for the group that received placebo to only 10% in the corticosteroid-treated groups. The results in the retrospective study were virtually identical. While the Schumer study overcame many of the criticisms raised by Weitzman and Berger, there are still some major reservations about the study design and the conclusions reached (Shine et al 1980). Focusing only on the prospective study (the retrospective analysis merely served to reinforce the conclusions of the prospective study), the use of two different corticosteroid preparations was puzzling since the antiinflammatory effects of dexamethasone are more prolonged than those of methylprednisolone. Since physicians had the option of repeated doses, uniform dosing in the treated group was not given. These are relatively minor objections, however, compared with the significant omission of any information about fluid or volume therapy or the use of sympathomimetic amines like dopamine or norepinephrine to support blood pressure, renal perfusion and urine output. Another major variable that might have affected outcome relates to the selection and dose of antimicrobial therapy.

If the effect of steroids is to be critically assessed, all other major variables affecting outcome should be eliminated or minimized. Corticosteroids should have a beneficial effect that further reduces mortality below that achieved by the 'best available' conventional therapy. The latter should include optimal antimicrobial therapy. The Schumer study involved 8 years prior to and including 1974. During the first 3 years of the study, chloramphenicol was used as the only agent for treatment of the underlying bacterial infection. For the remainder of the study, clindamycin plus gentamicin was the routine antibacterial regimen. A substantial number of the patients involved in the study had serious infections due to organisms like *Pseudomonas aeruginosa* that are resistant to chloramphenicol. The extraordinary success of corticosteroids in the face of inappropriate antibacterial therapy thus make the results even more puzzling. One might even come to the incredulous conclusion that corticosteroids could overcome or counterbalance the effects of inappropriate antimicrobial treatment! Even the combination of clindamycin and gentamicin does not constitute optimal treatment for many aerobic gram-negative bacillary infections, such as those caused by *P. aeruginosa*.

Another major problem in the analysis of this study lies in the attribution of the cause of death. When a patient died within 4 weeks of entry to study, such a subject was considered evaluable for the effect (or lack of effect) of placebo or study drug. In view of the significant underlying diseases of these patients (a number had leukemia), it is questionable at best to attribute a death 4 weeks later to septic shock. Still, all of the objections or criticisms voiced above do not explain the remarkable salvage rate in the patients receiving large doses of corticosteroids. The unusually low mortality rate in the steroid-treated group is perhaps the single major reservation that most clinicians have expressed upon reviewing this study. If it is true that 9 out of 10 patients meeting well-accepted criteria for shock will survive after large doses of intravenous steroids, such a clear-cut result should be easy to duplicate in the hands of others.

4. RECENT DEVELOPMENTS

Although they have recognized the limitations of the Schumer study, many clinicians believe that a short high-dose course of corticosteroids probably does no harm and they do not object to their use when other supportive measures seem to be failing. On the other hand, the cost of these medications is high and corticosteroids may further impair host defenses in an already immunosuppressed host. A more recent published review of 612 cases of gram-negative rod bacteremia included 269 (44%) cases of shock (Kreger et al 1980). Most of these patients were not treated with steroids, and there was no evidence that severity of shock influenced selection of steroid therapy. The highest fatality rates occurred in each category of underlying disease (by the McCabe and Jackson criteria) among patients who received more than 400 mg hydrocortisone per day, and these differences were statistically significant from those who did not receive corticosteroids. Admittedly, this was not a randomized prospective study using a consistent corticosteroid dosage or uniform antimicrobial protocol. The retrospective nature of analysis makes this report subject to many of the criticisms of Weitzman and Berger. Finally, the doses of corticosteroids used were relatively low in comparison to those advocated by Schumer.

Perhaps the most telling conclusion that can be drawn about all published studies to date is that upon exhaustive review of the clinical studies by the U.S. Food and Drug Administration, including the records of many of the patients studied, this regulatory body decided to remove septic shock as one of the approved indications for the therapeutic use of corticosteroids (Sheagren 1981).

5. OUTLOOK FOR THE FUTURE

The proponents of the use of so-called 'industrial doses' of corticosteroids have clung tenaciously to the hope that such forms of intervention might be beneficial. The therapeutic dose has been escalated, the recommendation has been made that they be given as early as possible, and the rationale offered that 1 or 2 large doses that are 'pulsed' can do little harm. It has also been argued that patients who meet classical criteria for septic shock have reached the point that is relatively refractory to intervention (Sheagren 1981). At that point, mortality exceeding 50% may prevail and any form of intervention may not further modify clinical outcome. Proponents of corticosteroid therapy have suggested that to have an effect, treatment should be aimed at inhibiting inflammatory systems at the very early stages of septic shock, perhaps even 'pre-shock'. The danger, of course, is that a very large number of patients would probably be over-treated if this overly aggressive philosophy were adopted.

In a recent symposium on septic shock, Sheagren suggested that 4 out of the 7 following criteria be met before initiating steroid therapy for the presumptive diagnosis of incipient septic shock (Sheagren 1985): a. altered mentation; b. orthostatic changes in blood pressure; c. tachypnea and decreased CO_2; d. decreased blood granulocytes; e. decreased blood platelet concentrations; f. serious cardiac, pulmonary or renal disease; g. hypothermia.

After an initial dose of 30 mg/kg as advocated by Schumer, a second dose of steroids within 4–6 hours was recommended if hemodynamic improvement did not occur. It was suggested that if shock persists despite normalization of volumes and pressures, an inotropic drug be given.

One might argue that it would be more prudent to give a fluid challenge and dopamine or dobutamine before resorting to a corticosteroid 'pulse'. Comprehensive antibacterial chemotherapy would clearly be administered at first diagnosis of infection.

Regardless of how one views the present situation, two randomized, double-blind clinical trials of high-dose corticosteroid therapy for human septic shock are currently underway in the U.S. One is sponsored by the U.S. Veterans Administration and another study involves multiple collaborating hospitals. Because of the many variables affecting clinical outcome, it would seem advisable to structure these studies in a manner that will determine what steroids will do in addition to appropriate antimicrobials, volume replacement, and agents such as the inotropic sympathomimetic amines. Since there appear to be major improvements in overall supportive care and antimicrobial therapy in the interim from the completion of the Schumer study, only a large number of cases entered into these well-designed trials and meticulous execution of study design will provide a convincing answer to the controversy of decades about the use of corticosteroids for septic shock in man.

REFERENCES

Gifford RH, Feinstein AR (1969) A critique of methodology in studies of anticoagulant therapy for acute myocardial infarction. *N. Engl. J. Med. 280*, 351-357.

Kimball HR, Melmon KL, Wolff SM (1972) Endotoxin induced kinin production in man. *Proc. Soc. Exp. Biol. Med. 139*, 1078-1082.

Klastersky J, Cappel R, Debusscher L (1971) Effectiveness of betamethasone in management of severe infections – a double blind study. *N. Engl. J. Med. 284*, 1248-1250.

Kreger BE, Crave DE, McCabe WR (1980) Gram-negative bacteremia. IV. Reevaluation of clinical features and treatment in 612 patients. *Am. J. Med. 68*, 344-355.

McCabe WR (1974) Gram-negative bacteremia. *Adv. Intern. Med. 19*, 135-158.

McCabe WR, Jackson GG (1962) Gram-negative bacteremia. II. Clinical laboratory and therapeutic observations. *Arch. Intern. Med. 110*, 856-864.

Schumer W (1976) Steroids in the treatment of clinical septic shock. *Ann. Surg. 184*, 333-341.

Sheagren JN (1981) Septic shock and corticosteroids. *N. Engl. J. Med. 305*, 456-458.

Sheagren JN (1985) Glucocorticoid therapy in the management of severe sepsis. In: Sande M, Root RK (Eds), *Symposium on Septic Shock*, pp 201-218. Churchill Livingstone, New York.

Shine K, Silver M, Young, LS, Tillisch J (1980) Aspects of the management of shock. *Ann. Intern. Med. 93*, 723-734.

Weitzman S, Berger S (1974) Clinical trial design in studies of corticosteroids for bacterial infections. *Ann. Intern. Med. 81*, 36-42.

Wolff S (1973) Biological effects of bacterial endotoxin in man. *J. Infect. Dis. (Suppl) 128*, S259-S263.

Young LS (1985) Gram-negative sepsis. In: Mandell G, Douglas RG Jr, Bennett JE (Eds), *Principles and Practice of Infectious Diseases*, pp 452-475. John Wiley, New York.

Subject index